UNIX

A HISTORY AND A MEMOIR

ALSO BY BRIAN KERNIGHAN

The Elements of Programming Style (with P. J. Plauger)

Software Tools (with P. J. Plauger)

Software Tools in Pascal (with P. J. Plauger)

The C Programming Language (with Dennis Ritchie)

The AWK Programming Language
(with Al Aho and Peter Weinberger)

The Unix Programming Environment (with Rob Pike)

AMPL: A Modeling Language for Mathematical Programming
(with Robert Fourer and David Gay)

The Practice of Programming (with Rob Pike)

D is for Digital

Hello, World: Opinion Columns from the Daily Princetonian

The Go Programming Language (with Alan Donovan)

Understanding the Digital World

Millions, Billions, Zillions

UNIX

A History and a Memoir

Brian Kernighan

Kindle Direct Publishing

Copyright © 2020 by Brian W. Kernighan

Published by Kindle Direct Publishing

www.kernighan.com

All Rights Reserved

ISBN 978-169597855-3

Camera-ready copy for this book was produced by the author in Times Roman and Helvetica, using groff, ghostscript, and other open source Unix tools.

Printed on acid-free paper. ∞

Printed in the United States of America

10 9 8 7 6 5 4 3 2 1

In memoriam DMR

Contents

Preface   ix

Chapter 1: Bell Labs   1
  1.1 Physical sciences at Bell Labs   5
  1.2 Communications and computer science   7
  1.3 BWK at BTL   8
  1.4 Office space   11
  1.5 137 → 127 → 1127 → 11276   19

Chapter 2: Proto-Unix (1969)   27
  2.1 A bit of technical background   27
  2.2 CTSS and Multics   30
  2.3 The origin of Unix   32
  2.4 What's in a name?   34
  2.5 Biography: Ken Thompson   35

Chapter 3: First Edition (1971)   41
  3.1 Unix for patent applications   42
  3.2 The Unix room   45
  3.3 The Unix Programmer's Manual   49
  3.4 A few words about memory   52
  3.5 Biography: Dennis Ritchie   55

Chapter 4: Sixth Edition (1975)   61
  4.1 File system   62
  4.2 System calls   63
  4.3 Shell   65
  4.4 Pipes   67
  4.5 Grep   70
  4.6 Regular expressions   73
  4.7 The C programming language   76
  4.8 Software Tools and Ratfor   80
  4.9 Biography: Doug McIlroy   82

Chapter 5: Seventh Edition (1976-1979) 87
    5.1 Bourne shell 88
    5.2 Yacc, Lex, Make 90
    5.3 Document preparation 98
    5.4 Sed and Awk 113
    5.5 Other languages 117
    5.6 Other contributions 121

Chapter 6: Beyond Research 131
    6.1 Programmer's Workbench 131
    6.2 University licenses 134
    6.3 User groups and Usenix 136
    6.4 John Lions' Commentary 137
    6.5 Portability 140

Chapter 7: Commercialization 143
    7.1 Divestiture 143
    7.2 USL and SVR4 144
    7.3 UNIX™ 146
    7.4 Public relations 147

Chapter 8: Descendants 153
    8.1 Berkeley Software Distribution 153
    8.2 Unix wars 156
    8.3 Minix and Linux 158
    8.4 Plan 9 160
    8.5 Diaspora 163

Chapter 9: Legacy 165
    9.1 Technical 166
    9.2 Organization 170
    9.3 Recognition 175
    9.4 Could history repeat? 177

Sources 181

Preface

"One of the comforting things about old memories is their tendency to take on a rosy glow. The memory fixes on what was good and what lasted, and on the joy of helping to create the improvements that made life better."
     Dennis Ritchie, "The Evolution of the Unix Time-sharing System," October 1984

Since its creation in a Bell Labs attic in 1969, the Unix operating system has spread far beyond anything that its creators could possibly have imagined. It has led to the development of much innovative software, influenced myriad programmers, and changed the entire path of computer technology.

Unix and its derivatives aren't widely known outside a particular technical community, but they are at the heart of any number of systems that are part of everyone's world. Google, Facebook, Amazon, and plenty of other services are powered by Linux, a Unix-like operating system that I'll talk about later on. If you have a cell phone or have a Mac, it runs some version of Unix. If you have gadgets like Alexa at home or navigation software in your car, they're powered by Unix-like systems too. If you're bombarded by advertising whenever you browse the web, Unix systems are behind it, and of course the tracking that knows what you're doing so you can be more accurately bombarded is likely to be based on Unix as well.

Unix was created more than 50 years ago by two people, along with a small group of collaborators and camp followers. Through a sequence of lucky accidents, I was present at the creation, though certainly not responsible for any of it. At most I can take credit for a modest amount of useful software and, thanks to first-rate co-authors, some books that have helped people to learn more about Unix and its languages, tools and philosophy.

This book is part history and part memoir, a look at the origins of Unix and an attempt to explain what Unix is, how it came about, and why it matters. The book is certainly not a scholarly work, however—there are no footnotes—and it has become less history and more memoir than I had originally planned.

The book is written for anyone with an interest in computing or the history of inventions. It includes a certain amount of technical material, but I've tried to provide sufficient explanation that even if you don't have much background, you can appreciate the basic ideas and see why they might be important. You can safely skip over anything that seems too complicated, however, so there's no need to read every word. If you are a programmer, some of the explanations may seem pretty obvious and over-simplified, but with luck some of the historical insights will be useful, and the stories that go with it might be new and interesting to you.

Although I have tried to be accurate, there are sure to be places where my memory is imperfect. Furthermore, the interviews, personal reminiscences, oral histories, books and papers that I've relied on are not always consistent with my own memories or with each other in their accounts of who did what and when.

Fortunately, many of those involved in the early days are still alive and have helped to straighten me out. They too may suffer from memory lapses and rose-colored glasses but any errors that remain are my fault, at least until I can safely blame them on someone else.

My main purpose in writing is to tell some of the wonderful stories of an especially productive and formative time in the history of computing. It's important to understand the evolution of the technology that we use and take for granted. The decisions that shaped how that technology developed and thus defined the paths that we took were made by real people, working under the pressures and constraints of the time. The more we know about the history, the more we can appreciate the inventive genius that led to Unix and perhaps better understand why modern computer systems are as they are. If nothing else, choices that might seem wrong-headed or perverse today can often be seen as natural consequences of what was understood and could be accomplished with the resources available at the time.

The story is about more than just the Unix operating system, though that's the core. It also includes the C programming language, one of the most widely used of all languages, and at the heart of the systems that run the Internet and the services that use it. Other languages began life at Bell Labs with Unix too, notably C++, which is also widely used. Microsoft Office tools like Word, Excel or Powerpoint are written in C++, as are most of the browsers you might be using. A dozen or two of the core tools that programmers use daily and simply take for granted were written in the early days of Unix and are still part of every programmer's toolkit, often much the same as they were 40 or 50 years ago.

Computer science theory plays a vital role as well, often enabling immensely practical tools. Hardware research explored design tools, integrated circuits, computer architecture, and unusual special-purpose devices. The interplay among all of these activities often led to unexpected inventions, and was one of the reasons why the whole enterprise was so productive across so many different fields.

There is also an interesting and relevant story about how technological innovation happens. Bell Labs, where Unix began, was a remarkable institution that produced many good ideas and capitalized on them. It was the origin of many world-changing inventions, and there are lessons to be learned from how it worked.

The Unix story certainly offers many insights into how to design and build software, and how to use computers effectively, which I have tried to highlight along the way. As a simple but characteristic example, the Unix philosophy of software tools made it possible to combine existing programs to accomplish a wide variety of tasks without having to write new software. It's a programming instance of an old strategy: divide and conquer. By breaking bigger tasks into smaller ones, each one becomes more manageable, and the pieces can be combined in unexpected ways.

Finally, although Unix was the most visible software from Bell Labs, it was by no means the only contribution to computing. The Computing Science Research Center, the fabled "Center 1127," or just "1127," was unusually productive for two or three decades. Its work was inspired by Unix and used Unix as a base, but the contributions go well beyond that. Members of 1127 wrote important books that for years have been core texts in computer science and references for programmers. Center 1127 was an exceptionally influential industrial computer science research laboratory, one of the most productive of comparably sized groups at the time or subsequently.

Why was Unix and the surrounding environment so successful? How did a two-person experiment grow into something that literally changed the world? Was this a singular event, so unlikely that nothing like it could ever happen again? On the larger question of whether such influential results can be planned, I'll say more at the end of the book. For now, it seems to me that Unix owes its success to an accidental combination of factors: two exceptional people, an excellent supporting cast, talented and enlightened management, stable funding in a corporation that took a very long view, and an unfettered environment for exploration no matter how unconventional. Its adoption was facilitated by rapidly advancing technology where hardware kept getting smaller, cheaper and faster at an exponential rate.

The early years of Unix were for me and many others at Bell Labs wondrously productive and fun. I hope that this book will help you sense some of the joy of creation, and indeed of making life better, that Dennis Ritchie described in the epigraph above.

Acknowledgments

One of the unexpected pleasures of writing this book has been to reconnect with so many friends and colleagues, who have generously shared their memories and some great stories. It is hard to express just how valuable that has been. I haven't been able to include all of the stories, but I have greatly enjoyed hearing them, and I am in debt to the many remarkable people that I have been fortunate enough to work with.

Biographical material is scattered throughout the book, with major sections on the three main people without whom Unix would not have happened—Ken Thompson, Dennis Ritchie and Doug McIlroy. Ken and Doug have provided invaluable feedback on the book, though they are in no way responsible for anything I've gotten wrong or inadvertently misrepresented. I have also received valuable comments and suggestions from Dennis's brothers John and Bill; his nephew Sam provided detailed comments on several drafts.

As he has done so many times before, Jon Bentley gave me invaluable insights, helpful suggestions for organization and emphasis, numerous anecdotes, and detailed comments on writing, over at least half a dozen drafts. I am enormously indebted to Jon once again.

Gerard Holzmann provided advice, archival material and many original photographs that have helped to make the book more visually interesting.

Paul Kernighan read multiple drafts and spotted myriad typos. He also suggested some excellent titles, though in the end I regretfully decided not to use *A History of the Unix-speaking Peoples*.

Al Aho, Mike Bianchi, Stu Feldman, Steve Johnson, Michael Lesk, John Linderman, John Mashey, Peter Neumann, Rob Pike, Howard Trickey and Peter Weinberger provided critical reading and stories of early Unix days, many of which I have quoted or paraphrased.

I also received many helpful comments and other contributions from Michael Bachand, David Brock, Grace Emlin, Maia Hamin, Bill Joy, Mark Kernighan, Meg Kernighan, William McGrath, Peter McIlroy, Arnold Robbins, Jonah Sinowitz, Bjarne Stroustrup, Warren Toomey and Janet Vertesi.

I am deeply grateful for all of the generous assistance, but I am responsible for any errors or misinterpretations. Many other people have contributed to Unix in important ways over the past fifty years, and I apologize to anyone whose work has been slighted.

Chapter 1

Bell Labs

"One Policy, One System, Universal Service"
    AT&T's mission statement, 1907

"At first sight, when one comes upon it in its surprisingly rural setting, the Bell Telephone Laboratories' main New Jersey site looks like a large and up-to-date factory, which in a sense it is. But it is a factory for ideas, and so its production lines are invisible."
    Arthur Clarke, *Voice Across the Sea*, 1974
        quoted in *The Idea Factory* by Jon Gertner, 2012

To understand how Unix happened, we have to understand Bell Labs, especially how it worked and the creative environment that it provided.

AT&T, the American Telephone and Telegraph Company, grew out of combining a host of local telephone companies from across the United States. Early in its history, AT&T realized that it needed a research organization that would systematically address the scientific and engineering problems that the company encountered as it tried to provide a national telephone system. In 1925, it created a research and development subsidiary, Bell Telephone Laboratories, to attack these problems. Although the full name was regularly abbreviated to Bell Labs or BTL or merely "the Labs," telephony was always the central concern.

Bell Labs was originally located at 463 West Street in New York City, but at the beginning of the Second World War, many of its activities moved out of New York. AT&T was heavily involved in the war effort, providing expertise on a wide variety of important military problems—communications systems, of course, but also fire-control computers for anti-aircraft guns, radar, and cryptography. Part of this work was done in suburban and rural New

Figure 1.1: From New York City to Murray Hill, New Jersey

Jersey, 20 miles (33 km) west of New York. The largest site was in an area called Murray Hill, which was part of the small towns of New Providence and Berkeley Heights.

Figure 1.1 shows the general lay of the land; 463 West Street is on the Hudson River, a short distance north of the 9A highway marker. Bell Labs at Murray Hill straddles the boundary between New Providence and Berkeley Heights, just north of Interstate 78. Both locations are marked on the map with dots.

More and more Bell Labs activities shifted to Murray Hill, and the Labs left West Street entirely in 1966. By the 1960s, Murray Hill housed over 3,000 people, at least 1,000 with PhDs in technical fields like physics, chemistry, mathematics, and various flavors of engineering.

Figure 1.2 is an aerial photo of the Murray Hill complex in 1961. There were three main buildings. Building 1 is to the lower right in the picture, 2 is to the upper left, and 3 is the square one with an open courtyard. Before it was blocked by the addition of two new buildings in the 1970s, there was a single uninterrupted quarter-mile (400 m) corridor from one end of Building 1 to the other end of Building 2.

I spent over 30 years in Building 2, from an internship in 1967 until I retired in 2000. My offices were in the side wings marked with dots, on the fifth (top) floor. For future reference, Stair 9 in this picture is at the absolute far end of Building 2, and Stair 8 is one wing closer to the center. For most of the early years, the Unix room was in the sixth floor attic between stairways 8 and 9.

Figure 1.3 shows a Google satellite image of Bell Labs in 2019. Buildings 6 (towards the lower left, with the marker) and 7 (towards the upper right) were added in the early 1970s, and for some years after 1996, Building 6 was the headquarters of Lucent Technologies. It's intriguing how much corporate history is captured in the labels that Google has assigned: "Bell Labs" as a marker, Lucent Bell Labs on the exit driveway, Alcatel-Lucent Bell Labs on the entry, and Nokia Bell Labs at the apex of the managerial pyramid in Building 6.

I'm not qualified to write a detailed history of the Labs, but fortunately that's already been done well by other writers. I particularly like Jon Gertner's *The Idea Factory*, which focuses on the physical sciences, and James Gleick's *The Information* is excellent for information science. The voluminous (seven volumes and nearly 5,000 pages) official Bell Labs publication called *A History of Science and Engineering in the Bell System* is thorough, authoritative, and in my sampling, always interesting.

Figure 1.2: Bell Labs in 1961 (Courtesy of Bell Labs)

1.1 Physical sciences at Bell Labs

During its early years, most research at Bell Labs involved physics, chemistry, materials, and communications systems. Researchers had exceptional freedom to pursue their own interests, but the environment was so rich in relevant problems that it wasn't hard to explore in areas that were both scientifically interesting and potentially useful to the Bell System and to the world at large.

Bell Labs was responsible for a remarkable number of scientific and technological advances that changed the world. Foremost among them was the transistor, invented in 1947 by John Bardeen, Walter Brattain and William Shockley, who were trying to improve amplifiers for long-distance telephone circuits. The transistor resulted from fundamental research into the properties of semiconductor materials, driven by a need for devices that would be more physically robust and less energy-hungry than vacuum tubes, which in the 1940s were the only way to make communications equipment and, incidentally, to build the earliest computers.

The invention of the transistor was recognized with a Nobel Prize in physics in 1956, one of nine Nobel prizes that have been awarded to scientists for work done at least in part at Bell Labs. Other major inventions included negative feedback amplifiers, solar cells, lasers, cell phones, communications satellites and charge-coupled devices (which make the camera in your phone work).

Very roughly, in the 1960s through 1980s there were 3,000 people in the research area of Bell Labs (mostly at Murray Hill), and 15,000 to 25,000 in development groups in multiple locations that designed equipment and systems for the Bell System, often using results from the research area. That's a lot of people. Who paid for them all?

AT&T was effectively a monopoly, since it provided telephone service to most of the United States, but its ability to exploit its monopoly power was constrained. It was regulated by federal and state bodies that controlled the prices that it could charge for its various services, and it was not allowed to enter businesses that were not directly related to providing telephone services.

This regulatory regime worked well for many years. AT&T was required to provide service to everyone ("universal service") no matter how remote or unprofitable. As compensation, it got a stable and predictable overall rate of return.

As part of this arrangement, AT&T directed a small fraction of its revenue to Bell Labs, with the express purpose of improving communications

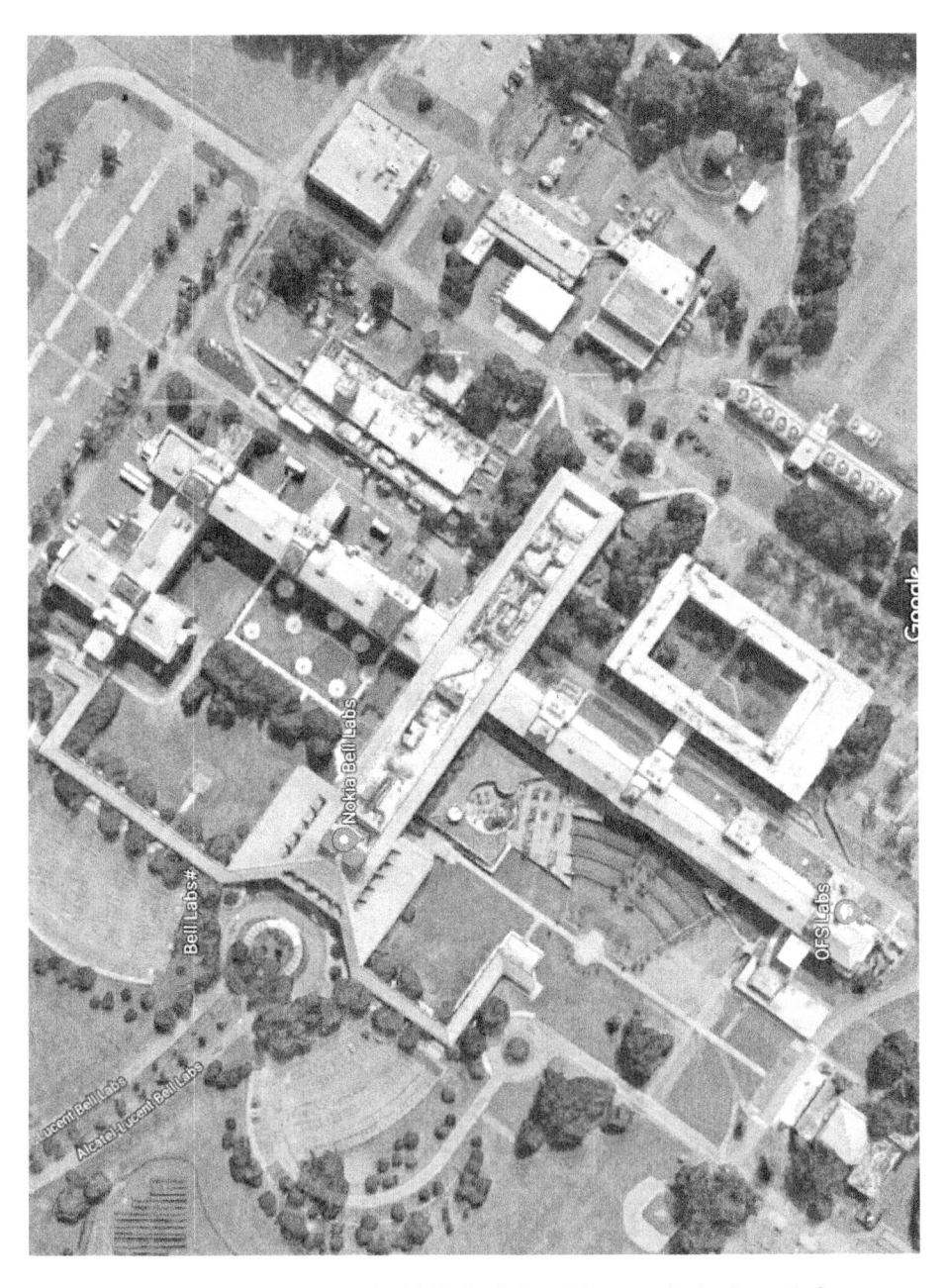

Figure 1.3: Bell Labs in 2019; Building 6 is towards the lower left

services. In effect, Bell Labs was paid for by a modest tax on every phone call in the country. According to a paper by A. Michael Noll, AT&T spent about 2.8 percent of its revenues on research and development, with about 0.3 percent on basic research. I'm not sure how well this would work today, but for decades, the arrangement led to a steady flow of improvements to the phone system and a significant number of fundamental scientific discoveries.

Stable funding was a crucial factor for research. It meant that AT&T could take a long-term view and Bell Labs researchers had the freedom to explore areas that might not have a near-term payoff and perhaps never would. That's a contrast with today's world, in which planning often seems to look ahead only a few months, and much effort is spent on speculating about financial results for the next quarter.

1.2 Communications and computer science

Bell Labs was naturally a pioneer in designing, building and improving communications systems, a blanket term that covered everything from the design of consumer hardware like telephones through to the infrastructure of switching systems, microwave transmission towers, and fiber optic cables.

Sometimes this breadth of practical concerns could even lead to advances in basic science. For example, in 1964 Arno Penzias and Robert Wilson were trying to figure out what was causing unwanted noise in the antenna that Bell Labs was using to detect radio signals bounced off Echo balloon satellites. They eventually deduced that the noise came from the background radiation that was the residue of the cosmic Big Bang at the beginning of the universe. This discovery led to the 1978 Nobel Prize in physics for Penzias and Wilson. (Arno says that "Most people get Nobels for things they were looking for. We got one for something we were trying to get rid of.")

Another part of the Bell Labs mission was to develop a deep mathematical understanding of how communications systems worked. The most important result was Claude Shannon's creation of information theory, which was in part motivated by his study of cryptography during World War II. His 1948 paper "A Mathematical Theory of Communication," published in the *Bell System Technical Journal*, explained the fundamental properties and limitations on how much information could be sent through a communications system. Shannon worked at Murray Hill from the early 1940s to 1956, then returned to teach at MIT, where he had been a graduate student. He died in 2001 at the age of 84.

As computers became more powerful and less expensive, computer use expanded to include more data analysis, along with extensive modeling and simulation of physical systems and processes. Bell Labs had been involved with computers and computing since the 1930s, and had computer centers with big central computers by the late 1950s.

In the early 1960s, a computer science research group was formed by splitting some people out of mathematics research, along with some of the staff who operated the large central computer at Murray Hill. The resulting amalgam was called the Computing Science Research Center, and although for a brief period it still ran computer services for all of Murray Hill, it was part of the research division, not a service function. In 1970 the group that managed the computer facilities was moved into a separate organization.

1.3 BWK at BTL

This section contains a fair amount of personal history, which I hope will give you some idea of the lucky accidents that led me to computing as a career, and to Bell Labs as an unequaled place to pursue it.

I was born in Toronto, and went to the University of Toronto. I was in a program called Engineering Physics (later renamed to Engineering Science), a catch-all program for those who didn't really know what they wanted to focus on. I graduated in 1964, which was in the early days of computing: I saw my first computer when I was in my third year at university. There was only one big computer for the whole university, an IBM 7094, which was pretty much top of the line. It had 32K (32,768) 36-bit words of magnetic core memory (today we would say 128K bytes), and some secondary storage in the form of big mechanical disk drives. It cost literally three million US dollars at the time and it lived in a large air-conditioned room, tended by professional operators; ordinary people (and especially students) did not get anywhere near it.

As a result, I did little computing as an undergraduate, though I did try to learn the Fortran programming language. For anyone who has ever struggled to write their first program, I can sympathize. I studied Daniel McCracken's excellent book on Fortran II and had the rules down pat, but I couldn't figure out how to write that first program, a conceptual barrier that many people seem to encounter.

In the summer before my final year of college, I landed a job at Imperial Oil in Toronto, in a group that developed optimization software for refineries. (Imperial Oil was partly owned by Standard Oil of New Jersey, which became Exxon in 1972.)

In retrospect, I was well below average as an intern. I spent the entire summer trying to write a giant Cobol program for analyzing refinery data. I don't recall its precise purpose, but I know for sure that it never worked. I didn't really know how to program, Cobol provided little support for good program organization, and structured programming had not been invented, so my code was an endless series of IF statements that branched off somewhere to do something once I had figured out what that something should be.

I also tried to get Fortran programs running on Imperial's IBM 7010, since I sort of knew Fortran, certainly better than I knew Cobol, and Fortran would have been better suited for data analysis. It was only after weeks of fighting JCL, IBM's Job Control Language, that I deduced that there was no Fortran compiler on the 7010, but the JCL error messages were so inscrutable that no one else had figured that out either.

When I returned to school for my senior year after this somewhat frustrating summer, I was still strongly interested in computing. There were no formal courses on computer science, but I did write my senior thesis on artificial intelligence, a hot topic at the time. Theorem provers, programs to play chess and checkers, and machine translation of natural languages all seemed like they were within reach, just requiring a bit of programming.

After graduating in 1964, I had no clue what to do next, so like many students I put off the decision by going to graduate school. I applied to half a dozen schools in the United States (not common among Canadians at the time), and by good luck was accepted by several, including MIT and Princeton. Princeton said that the normal time to complete a PhD was three years, while MIT said it would probably take seven years. Princeton offered a full fellowship; MIT said I would have to be a research assistant for 30 hours a week. The decision seemed pretty clear-cut, and a good friend, Al Aho, who had been a year ahead of me at Toronto, was already at Princeton, so off I went. It turned out to be an incredibly fortunate choice.

In 1966, I got lucky again, with a summer internship at MIT, thanks in part to the fact that another Princeton grad student, Lee Varian, had done a great job there in 1965. I spent the summer using CTSS, the Compatible Time-Sharing System, writing programs in MAD (Michigan Algorithm Decoder, a dialect of Algol 58) to build tools for a new operating system called Multics, which we'll come to in Chapter 2. (Multics was originally spelled MULTICS, but the lower-case version is less visually jarring; as with UNIX versus Unix and some other all-caps words, I'll use the nicer-looking form even though it's not historically accurate.)

My nominal boss at MIT was Professor Fernando Corbató, "Corby" to everyone, a wonderful gentleman, the creator of CTSS, and the person in charge of Multics. Corby won the Turing Award in 1990 for his fundamental work on time-sharing systems. He died in July 2019 at the age of 93.

In addition to leading the design and implementation of CTSS and Multics, Corby was the inventor of passwords for computer access. There had been little need for such a thing with batch computing, but some security mechanism was necessary to protect private files on a time-sharing computer with a shared file system.

I still remember both of the short pronounceable passwords that were automatically generated for me while I was at MIT in the summer of 1966. Two passwords were needed, not one, because one day during the summer a mixup caused the file that stored the passwords to be exchanged with the file that held the login message of the day. Since passwords were stored unencrypted, everyone who logged in got to see everyone else's passwords.

My office, which I remember as being unshared, belonged to Professor Joseph Weizenbaum, but he was away for the summer. Early in 1966, Weizenbaum had published a description of Eliza, a program that simulated a dialog with a Rogerian psychologist. Eliza was one of the first attempts at using natural language to interact with a computer. It was surprisingly effective, at least when first encountered, and I enjoyed showing it off to visitors. There are plenty of Eliza simulators on the web today; it's still entertaining to play with them for a few minutes, though their weaknesses quickly become evident.

```
ELIZA: Please tell me what's been bothering you.
YOU:   I'm trying to write a book about Unix.
ELIZA: Is it because you are trying to write a book
       about unix that you came to me?
YOU:   Yes.
ELIZA: You are sure.
YOU:   Yes.
ELIZA: You seem to be quite positive.
YOU:   Yes.
ELIZA: I see.
```

In the summer of 1967, I got the ultimate piece of good luck: an internship at Bell Labs in Murray Hill, in the Computing Science Research Center, working for Doug McIlroy (Figure 1.4). Doug suggested that I explore some problem in evaluating memory allocators, one of his long-term interests. In the best intern tradition, I bumbled around and eventually did something completely different, creating a library of functions that made it convenient to do list processing in Fortran programs. I spent the summer writing tight assembly language for the then-current big computer at Murray Hill, a GE 635, which was in effect a cleaned-up and more orderly IBM 7094, but also a simpler version of the GE 645 that had been specially designed for Multics. That's pretty much the last time I wrote assembly language, but even though what I was doing was fundamentally misguided, it was a blast and it hooked me completely on programming.

Figure 1.4: Doug McIlroy, ~1984 (Courtesy of Gerard Holzmann)

1.4 Office space

Sometimes geography is destiny.

My office as an intern in 1967 was on the fifth floor of Building 2, on a corridor off Stair 8. On my first day on the job, I was sitting in my office (the good old days when even an intern could luck into a private office), wondering what to do, when an older guy appeared in my doorway at 11AM, and said "Hi, I'm Dick [unintelligible]. Let's go to lunch."

OK, I thought, why not? I don't remember anything at all about the lunch, but I do remember that afterwards, Dick [unintelligible] went off somewhere else, and I sneaked along the corridor to read the name tag on his door. Richard Hamming! My friendly next door neighbor was famous, the inventor of error-correcting codes, and the author of the textbook for a numerical analysis course that I had just taken.

Dick (Figure 1.5) became a good friend. He was a man of strong opinions and not afraid to express them, which I think put off some people, but I enjoyed his company and over the years profited a great deal from his advice.

He was a department head, but there were no people in his department, which seemed odd. He told me that he had worked hard to achieve this combination of suitable title without responsibility, something that I came to appreciate only much later when I became a department head with a dozen people in my department.

Figure 1.5: Dick Hamming, ~1975, in his trademark plaid jacket (Wikipedia)

I was there in the summer of 1968 when he learned that he had won the ACM Turing Award, which today is considered the computer science equivalent of a Nobel prize. Dick's sardonic reaction: since the Nobel prize was at that time was worth $100,000 and the Turing award was worth $2,000, he said that he had won 2 percent of a Nobel prize. (This was the third Turing award; the first two went to Alan Perlis and Maurice Wilkes, also pioneers of computing.) Dick was cited for his work on numerical methods, automatic coding systems, and error-detecting and error-correcting codes.

Dick was the person who started me writing books, which has turned out to be a good thing. He had a fairly low opinion of most programmers, who he felt were poorly trained if at all. I can still hear him saying

> "We give them a dictionary and grammar rules, and we say, 'Kid, you're now a great programmer.'"

He felt that programming should be taught as writing was taught. There should be a notion of style that separated poor code from good code, and programmers should be taught how to write well and appreciate good style.

He and I disagreed on how this might be accomplished, but his idea was sound, and it led directly to my first book, *The Elements of Programming Style*, which I published in 1974 with P. J. "Bill" Plauger, who was at the time in the adjacent office. Bill and I emulated Strunk and White's *The*

Elements of Style by showing a sequence of poorly written examples, and explaining how to improve each of them.

Our first example came from a book that Dick had showed me. He came into my office one day carrying a numerical analysis text, all up in arms about how bad the numerical parts were. I saw only an awful chunk of Fortran:

```
      DO 14 I=1,N
      DO 14 J=1,N
   14 V(I,J)=(I/J)*(J/I)
```

If you're not a Fortran programmer, let me explain. The code consists of two nested DO loops, both ending at the line which has the label 14. Each loop steps its index variable from the lower limit to the upper limit, so in the outer loop I steps from 1 to N, and for each value of I, in the inner loop J steps from 1 to N. The variable V is an array of N rows and N columns; I loops over the rows and for each row, J loops over the columns.

This specific pair of loops thus creates an N by N matrix with 1's on its diagonal and 0's everywhere else, like this when N is 5:

```
1 0 0 0 0
0 1 0 0 0
0 0 1 0 0
0 0 0 1 0
0 0 0 0 1
```

The code relies on the fact that integer division in Fortran truncates any fractional part, so if I is not equal to J, one of the divisions will produce 0, but if I equals J (as it does on the diagonal), the result will be 1.

This all seemed way too clever to me, and misplaced cleverness is a bad idea in programming.

Rewriting it in a straightforward and obvious way leads to a clearer version: each time through the outer loop, the inner loop sets every element of row I to 0, and then the outer loop sets the diagonal element V(I,I) to 1:

```
C     MAKE V AN IDENTITY MATRIX
      DO 14 I = 1,N
         DO 12 J = 1,N
   12       V(I,J) = 0.0
   14    V(I,I) = 1.0
```

This also led to our first rule of programming style:

> *Write clearly—don't be too clever.*

Dick retired from Bell Labs in 1976 and went to the Naval Postgraduate School in Monterey, California, where he taught until his death early in 1998 at the age of 82. The story goes that one of his courses there was known to students as "Hamming on Hamming," which suggests an awkward parallel with this section of the book.

Dick thought hard all the time about what he was doing and why. He often said that "The purpose of computing is insight, not numbers," and he even had a tie with that written on it (in Chinese). One of his early insights was that computing would come to account for half the work at Bell Labs. None of his colleagues agreed, but in fact his estimate soon proved too conservative. He used to say that Friday afternoons were for thinking great thoughts, so he sat back and thought, though he was always welcoming to visitors like me at any time.

A few years after his retirement, Dick gave an insightful talk that distilled his advice on how to have a successful career, called "You and Your Research," which you can find on the web. He gave the first version of that talk at Bellcore in March 1986; Ken Thompson drove me there so we could hear it. I've been recommending the talk to students for decades—it really is worthwhile to read the transcript, or to watch one of the video versions.

Right across the hall from me in the summer of 1967 was Vic Vyssotsky (Figure 1.6), another incredibly smart and talented programmer. Vic was in charge of the Bell Labs part of Multics, a partner with Corby, but he still managed to find time to talk almost daily to a lowly intern. Vic pressed me into teaching a Fortran class to physicists and chemists who needed to learn how to program. The experience of teaching non-programmers turned out to be good fun. It got me over any fear of public speaking and made it easy to get into a variety of teaching gigs later on.

Shortly afterwards, Vic moved to another Bell Labs location where he worked on the Safeguard anti-missile defense system. He eventually returned to Murray Hill and became the executive director responsible for computer science research, and thus was my boss a couple of levels up.

By the spring of 1968, I had started to work on a problem for my PhD thesis, one suggested by my advisor, Peter Weiner. The problem was called *graph partitioning*: given a set of nodes connected by edges, find a way to separate the nodes into two groups of equal size such that the number of edges that connect a node in one group to a node in the other group is as small as possible. Figure 1.7 shows an example: any other partition of the nodes into two groups of five requires more than two edges between the two sets.

CHAPTER 1: BELL LABS   15

Figure 1.6: Vic Vyssotsky, ~1982 (Courtesy of Bell Labs)

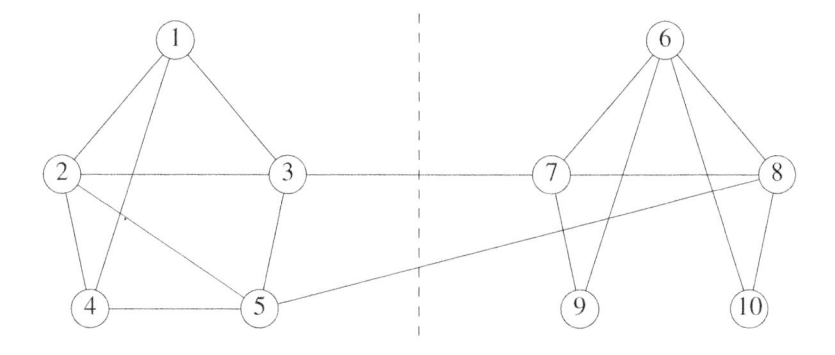

Figure 1.7: Graph partitioning example

This was ostensibly based on a practical problem: how to assign parts of a program to memory pages such that when the program was run, the amount of swapping of program pages into and out of memory would be minimized. The nodes represented blocks of code, the edges represented possible transitions from one block to another, and each edge could have a weight that measured the frequency of the transition and thus how costly it would be if the two blocks were in different pages.

It was an artificial problem in a way, but it was a plausible abstraction of something real, and there were other concrete problems that shared this abstract model. For example, how should components be laid out on circuit boards to minimize the expensive wiring that connected one circuit board to another? Less plausibly, how might we assign employees to floors of a building to keep people on the same floor as the people they talk to most often?

This was enough justification for a PhD thesis topic, but I wasn't making much progress. When I returned to the Labs for a second internship in the summer of 1968, I described the problem to Shen Lin (Figure 1.8), who had recently developed the most effective known algorithm for the classic Traveling Salesman problem: given a set of cities, find the shortest route that visits each city exactly once, and then returns home.

Shen came up with an approach for graph partitioning that looked promising, though there was no assurance that it would produce the best possible answers, and I figured out how to implement it efficiently. I did experiments on a large number of graphs to assess how well the algorithm worked in practice. It seemed highly effective, but we never discovered a way to guarantee an optimum solution. I also found a couple of interesting special-case graphs where I could devise algorithms that were both fast and guaranteed to produce optimal solutions. The combination of results was enough for a thesis, and by the end of the summer I had everything I needed. I wrote it up over the fall, and had my final oral exam late in January 1969. (Princeton's optimistic estimate of three years had turned into four and a half.)

A week later, I started work in the Computing Science Research Center at Bell Labs. I never had an interview; the Labs sent me an offer sometime in the fall, though with a caveat: my thesis had to be finished. Sam Morgan, the director of the Center and thus my boss two levels up, told me, "We don't hire PhD dropouts." Finishing the thesis was definitely a good thing; I got another letter in December saying that I had received a substantial raise, well before I reported for work!

As an aside, although Shen and I didn't know it at the time, there was a reason why we were unable to find an efficient graph-partitioning algorithm that would always find the best possible answer. Others had been puzzling over the inherent difficulty of combinatorial optimization problems like graph partitioning, and had discovered some interesting general relationships.

In a remarkable 1971 result, Stephen Cook, a mathematician and computer scientist at the University of Toronto, showed that many of these challenging problems, including graph partitioning, are equivalent, in the sense that if we could find an efficient algorithm (that is, something better than trying all possible solutions) for one of them, that would enable us to find efficient algorithms for all of them. It's still an open problem in computer science whether such problems are truly hard, but the betting is that they are. Cook received the 1982 Turing Award for this work.

When I got to Bell Labs as a permanent employee in 1969, no one told me what I should work on. This was standard practice: people were introduced to other people, encouraged to wander around, and left to find their own

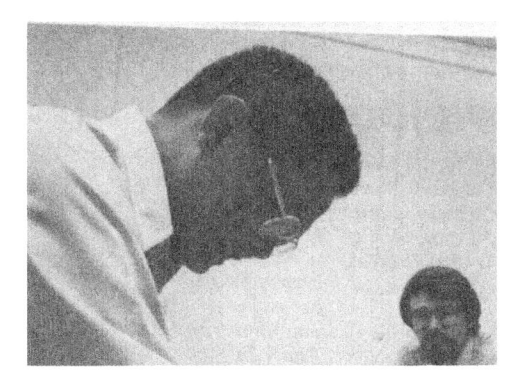

Figure 1.8: Shen Lin, ~1970 (Courtesy of Bell Labs)

research topics and collaborators. In retrospect, this seems like it must have been daunting, but I don't recall any concern on my part. There was so much going on that it wasn't hard to find something to explore or someone to work with, and after two summers I already knew people and some of the current projects.

This lack of explicit management direction was standard practice. Projects in 1127 were not assigned by management, but grew from the bottom up, coalescing a group of people who were interested in a topic. The same was true for work with other parts of the Labs: if I was involved with some development group, I might try to entice research colleagues to join me, but they would be volunteers.

In any case, for a while I continued to work with Shen on combinatorial optimization. Shen was exceptionally insightful on such problems, able to sense a promising line of attack by playing with small examples by hand. He had a new idea for the traveling salesman problem, a technique that greatly improved on his previous algorithm (which was already the best known), and I implemented it in a Fortran program. It worked well, and for many years was the state of the art.

This kind of work was fun and rewarding, but although I could convert ideas into working code pretty well, I wasn't any good at the algorithmic parts. So I gradually drifted into other areas: document preparation software, specialized programming languages, and a bit of writing.

I did come back to work with Shen a couple of other times, including a complex tool for optimizing the design of private networks for AT&T customers. It was good to flip back and forth between comparatively pure computer science and systems that were actually of some use to the company.

Brian W. Kernighan (co-author, *Partitioning Graphs*) is a member of the Computer Systems Research Department. He came to Bell Laboratories in February, 1969, and has been primarily interested in applications of graph models to computer programming and circuit layout problems.

Mr. Kernighan received the B.A.Sc. degree from the University of Toronto in 1964, and the Ph.D. degree from Princeton University in the computer science program in 1969. He is a member of the Association for Computing Machinery.

Brian W. Kernighan

Figure 1.9: PR photo, ~1970 (Courtesy of Bell Labs)

The Bell Labs public relations operation was fond of Shen's work on the Traveling Salesman problem and he figured in a number of advertisements. Figure 1.8 is a blurry excerpt from one of them, with me in the corner, and Figure 1.9, from some glossy PR magazine published by the Labs, talks about our work on graph partitioning, perhaps after we obtained a patent for the algorithm.

Note that, most uncharacteristically, I am wearing a tie. A few years later, Dennis Ritchie and I wrote an article about C for another company magazine, probably the *Western Electric Engineer*. Before publication, we were asked to send pictures of ourselves to accompany the article, which we did. After a few weeks, we were told that the pictures had been lost. No problem, we said, we can send them again. To which the response was "This time, could you wear ties?" We replied with a firm no, and shortly afterwards the magazine was published with our original tie-less pictures, which had miraculously been found.

When I started as a permanent employee, my office was on the fifth floor of Building 2 on a corridor off Stair 9, and I stayed in it for 30 years, a fixed point in a world of change. Over the years, my neighbors across the hall included Ken Thompson, Dennis Ritchie, Bob Morris, Joe Ossanna, and Gerard Holzmann, and eminent visitors like John Lions, Andy Tanenbaum and David Wheeler.

For the last decade of my time at the Labs, Ken Thompson and Dennis Ritchie's offices were directly across the corridor from mine. Figure 1.10 is a view of Dennis's office, taken in October 2005 from my old office doorway. Ken's office was to the left.

Over the years my immediate neighbors included Bill Plauger, Lorinda Cherry, Peter Weinberger and Al Aho. Doug McIlroy, Rob Pike and Jon

Figure 1.10: Dennis Ritchie's office in 2005

Bentley were just a few doors away. It's easier to collaborate with people who are physically close; I've been truly lucky in my neighbors.

1.5 137 → 127 → 1127 → 11276

Who were the players at this time, and what was the environment like? In the early 1970s, there were just over 30 people in Computing Science Research, with perhaps 4 to 6 working on Unix or things closely related to it. Figure 1.11 is a montage of part of the Bell Labs internal phone book. It's not yellowed with age; when I arrived, the org chart part was printed on yellow paper, just like the yellow pages in old telephone books.

The page is from 1969. It shows the Computing Science Research Center under Sam Morgan (Figure 1.12), who was an excellent applied mathematician and an expert in communications theory. Doug McIlroy, who played an enormously important but not widely known role in Unix, managed a group that included Ken Thompson and others who were involved in early Unix, like Rudd Canaday, Bob Morris, Peter Neumann and Joe Ossanna. Elliot Pinson's department included Dennis Ritchie, Sandy Fraser and Steve

	Computing Science Research Center			1373	Pinson E N, Head, Computer Systems Research Department	MH 2582
137	Morgan S P, Director, Computing Science Research Center	MH 6490			Blejwas Miss V M, Secretary	MH 2583
	Kalainikas Miss E, Secretary	MH 6491			Fraser A G	MH 3685
					Johnson S C	MH 3968
1371	McIlroy M D, Head, Computing Techniques Research Department	MH 6050			Kernighan B W	MH 6021
					Ritchie D M	MH 3770
					Sturman J N	MH 3164
	Marky Miss G A, Secretary	MH 6051			Winikoff A W	MH 2661
	Dimino L A	MH 2390		1374	Brown W S, Head, Computing Mathematics Research Department	MH 4822
	Aho A V	MH 4862				
	Canaday R H	MH 3038				
	Friedman A D	MH 4716			Blejwas Miss V M, Secretary	MH 4823
	Jensen P D	MH 6292			Hall A D	MH 4006
	Knowlton K C	MH 2328			Goldstein A J, Supervisor, Mathematical Techniques Group	MH 2655
	Menon P R	MH 2736				
	Morris R	MH 3878				
	Neumann P G	MH 2666			Lin S	MH 2111
	Ossanna J F	MH 3520			Shafer D M	MH 6862
	Thompson K L	MH 2394		1374	Traub J F, Supervisor, Numerical Mathematics Group	MH 2383
	Ullman J D	MH 6627				
	Wagner Mrs M R	MH 2879				
	Weiss Miss R A	MH 2007			Businger P A	MH 2059
					Richman P L	MH 3932
					Schryer N L	MH 2912
				1376	Hamming R W, Head, Computing Science Research Department	MH 2064
					Marky Miss G A, Secretary	MH 2065

Figure 1.11: Bell Labs phone book, ~1969 (Courtesy of Gerard Holzmann)

Johnson, who were also part of Unix for many years.

Although most researchers had PhDs, no one used "Doctor"; it was first names for everyone. One visible exception about titles was that in the phone book of Figure 1.11, women were either Miss or Mrs, while men were free of marital status indicators. I don't recall exactly when this labeling stopped, but it was certainly gone from the phone book by the early 1980s.

In the 1960s and 1970s there were few women and people of color in technical positions at Bell Labs; most members of technical staff were white males, and it stayed that way for a long time. In this respect, the Labs was representative of most technical environments at this period in the history of computing.

During the early 1970s, Bell Labs started three long-running programs that attempted to improve the situation. The Cooperative Research Fellowship Program (CRFP) began in 1972; each year, it funded four years or more of graduate school for about 10 minority students to obtain their PhDs. The Graduate Research Program for Women (GRPW), which started in 1974, provided the same graduate school support for women, perhaps 15 or 20 per year. Several of them worked in Center 1127 and in my department at one

Figure 1.12: Sam Morgan, director of 1127, ~1984 (Courtesy of Gerard Holzmann)

time or another, and most went on to successful careers within Bell Labs, at universities and at other companies. Each year, the Summer Research Program (SRP), also started in 1974, provided fully funded summer internships for roughly 60 undergraduate women and minority students, who were hosted at Murray Hill, Holmdel and sometimes other locations, working one-on-one with a research mentor. I was in charge of SRP for Center 1127 for over 15 years, so I got to meet a lot of nice sharp undergrads, and mentor a few.

These programs were effective in the long run, but the environment was still quite homogeneous during the 1960s and 1970s, and I'm sure that I was oblivious to some of the issues that this raised.

Bell Labs had a clear managerial hierarchy. At the top was the president, in charge of perhaps 15 to 25 thousand people. Beneath that were areas numbered 10 (research), 20 (development), 50 (telephone switching), 60 (military systems), and so on, each with a vice president. Research itself was divided into physics (11), mathematics and communications systems (13), chemistry (15) and the like, with an executive director for each; it also included legal and patent groups. Mathematics Research was 131, and Computing Science Research was "Center 137," with half a dozen individual departments like 1371. In a major shift, all of this was renumbered a few years later, when we became Center 127, and then during some reorganization, an extra digit was added to the front, and we became 1127, a number that lasted until 2005, well after I retired in 2000.

There were relatively few levels in the hierarchy. Researchers like me were "member of technical staff" or MTS, and there were a couple of technical levels below that. MTS in Research normally got a private office, though everyone was expected to keep the door open most of the time. There was a supervisor level above this, though 1127 had only a few supervisors over the years. The next level was department head, a person like Doug McIlroy who was responsible for half a dozen to a dozen individual researchers. The level above that was director of a center, who might have half a dozen departments, then executive director, with a handful of centers, and then vice president, who oversaw the executive directors.

Vice presidents reported to the president. Bill Baker, an outstanding chemist, was vice president of research from 1955 until 1973, and president of Bell Labs until 1980. While he was vice president, it was believed that he knew every MTS in research by name and was aware of what they were working on. I think it might well have been true; certainly he always knew what my colleagues and I were up to.

I was a regular MTS until 1981, when I finally succumbed to the pressure to become a department head. Most people went into management reluctantly, because although it was not the end of personal research, it did represent a slowdown, and it came with responsibilities like looking after one's department that could be challenging. Of course the usual arguments were trotted out: "It's inevitable, why not now?" Or, somewhat contradictory, "This might be your last chance." Or, "If not you, it will be someone else not as good."

For better or worse, I became head of a new department, 11276, with the carefully meaningless name "Computing Structures Research." The department usually had 8 to 10 people with a daunting spectrum of interests: graphics hardware, integrated circuit design tools, document preparation, operating systems, networking, compilers, C++, wireless system design, computational geometry, graph theory, algorithmic complexity, and lots more besides. Understanding what each of them was working on well enough to explain it further up the chain was always a challenge, though it was also rewarding, and a surprising amount of what I learned then has stuck with me.

The management hierarchy was accompanied by a few perks at each level. Some were obvious, like successively bigger offices at director and above. I think there was also a modest salary increase for department heads but it certainly wasn't big enough to be memorable.

Some were more subtle: department heads and up had carpeted offices, while ordinary folk had bare linoleum or vinyl tile. When I was promoted, I was given a glossy printed brochure listing my options for carpet color, office furniture, and the like. I briefly tried a new desk but it was too big and uncomfortable, so I went back to the ancient Steelcase that I had inherited in 1969. And I declined the carpet entirely, since I wasn't enthusiastic about distinctions of rank. Sam Morgan advised me strongly that I should take the carpet. He said that some day I would want the authority that came from having a carpet. I still declined, and the carpet distinction eventually went away.

The primary annual task for department heads was to assess the work of their department members in an elaborate ritual called "merit review." Once a year each MTS wrote down on one side of one piece of paper a summary of what they had done during the year; in 1127, it was known as an "I am great report," a term that I think originated with Sam Morgan. The department head wrote another piece of paper that summarized and assessed the work, including "areas for improvement," a section that was meant to contain constructive criticism.

Writing the assessment and feedback was hard work, and there was a strong tendency to leave the areas for improvement part blank, but one year we were told that it had to be filled in; evasions like leaving it empty or saying "N/A" were no longer acceptable. I came up with the phrase "Keep up the good work," and got away with that for a year or two before being told that more critical comments were required, on the grounds that no one was perfect. Fortunately I didn't have to do this for a star like Ken Thompson. What could one have said?

The department heads and the director met to come up with a consensus evaluation for each MTS. This normally took a full-day meeting. It was followed some weeks later by another full-day meeting that determined next year's salaries by allocating each MTS a share of a pot of raise money. These two related evaluations were officially known as merit review and salary review, but I always thought of them as "abstract merit" and "concrete merit."

This process was repeated up the management chain, with an executive director reviewing all the MTS results with directors, and also assessing department heads.

Although merit review in some centers could be competitive, our reviews were remarkably collegial. Rather than "My people are better than your people," the tone was more like "Don't forget this other good work that your person did."

I may be too sanguine, but I think the whole process worked well, because management was technically competent all the way up and everyone had gone through the process at lower levels. The system did not seem to have much of a bias towards either practice or theory, at least for us in 1127—good programs and good papers were both valued. The absence of proposals or plans for future work was a good thing. One was expected to have roughly a year's worth of accomplishments at the end of the year, but any number of false starts could be ignored, and management took a long view of people who worked on the same thing for multiple years. I think that it also helped that in Research there were very few management levels, so promotion wasn't really on most people's radar most of the time. If one really aspired to be a manager, an organization outside of Research would likely be a better option.

It's interesting to compare the Bell Labs evaluation process with how it's done in research universities. In the latter, hiring and especially promotion are strongly influenced by a dozen or more letters solicited from prominent outside researchers who are in the same specialty. This tends to encourage deep expertise in narrow fields, since the goal of a reviewee is to master some field so well that external reviewers can legitimately say "This person is the best person in this sub-field at this stage of his or her career."

By contrast, Bell Labs created a rank order for every researcher, from the bottom up. Each department head ranked his or her people; those rankings were merged by department heads within a center, and those in turn by the next two levels, so by the end everyone's approximate position in the whole population was determined.

A person who did great work in a narrow field might well be ranked highly by his or her immediate management, but no one further up would likely know of the work. Interdisciplinary work, on the other hand, stood out at higher levels because more managers would have seen it. The broader the collaborations, the more managers would know about it. The end result was an organization that strongly favored collaboration and interdisciplinary research. And because the managers who made the decisions had come up through the same process, they were inclined in the same direction.

I was a department head for over 15 years, was probably at best an average manager, and was definitely happy to step down. Others successfully resisted promotion for long periods; Dennis Ritchie became a department head well after I did, and Ken Thompson never did.

Having now taught at a university for 20 years, I'm still not enthusiastic about sitting in judgment on other people's work. It's necessary, however, and sometimes one has to make decisions that do affect people's lives, like firing someone (which I fortunately never had to do) or failing a student in a course (not common but not unheard of either). One of the good things about the Bell Labs process was that it was based on the shared judgment of other people who understood the work. As Doug McIlroy said, "Collegiality was the genius of the system. Nobody's advancement depended on the relationship with just one boss." The process at the Labs wasn't perfect, but it was pretty good and I've certainly heard and read about performance review processes that are far worse.

Chapter 2

Proto-Unix (1969)

"At some point I realized that I was three weeks from an operating system."

Ken Thompson, Vintage Computer Festival East, May 4, 2019

The Unix operating system was born in 1969, but it didn't spring into existence out of nothing. It came out of the experiences of several people at Bell Labs who had worked on other operating systems and languages. This chapter tells that story.

2.1 A bit of technical background

This section is a short primer on the basic technical material that forms the central topics of the book: computers, hardware, software, operating systems, programming, and programming languages. If you're already familiar with these ideas, skip ahead; if not, I hope this will bring you up to speed enough that you can appreciate what it's all about. If you want more detailed explanations aimed at non-technical readers, you might like my *Understanding the Digital World*, though I am not unbiased.

A computer is fundamentally not much more than a calculator like the ones that used to be separate gadgets but are now just applications on a phone. Computers can do arithmetic computations exceedingly fast, however, billions per second today, though it was significantly less than millions per second in the 1970s.

A typical computer of the 1960s and 1970s had a repertoire of a few dozen kinds of instructions that it could perform: arithmetic (add, subtract, multiply, divide), read information from primary memory, store information into primary memory, and communicate with devices like disks and anything else that might be connected. Plus one other crucial thing: the repertoire includes instructions that decide which instructions to perform next—thus what the computer will do next—based on the results of previous computations, that is, what it has already done. In that way, a computer controls its own destiny.

Instructions and data are stored in the same primary memory, which is usually called RAM, for "random access memory." If you load a different set of instructions into the RAM, the computer does a different job when it executes them. That's what's happening when you click on an icon for a program like Word or the Chrome browser—it tells the operating system to load the instructions for that program into memory and start to run it.

Programming is the process of creating the sequences of operations that perform some desired task, using some programming language. It's possible to create the necessary instructions directly, but this is a difficult chore with many clerical details, even for tiny programs, so most of the advances in programming have involved creating programming languages that are closer to the way that humans might express a computation. Programs called *compilers* (which of course have to be written themselves) translate from higher-level languages (closer to human language) ultimately to the individual instructions of a specific kind of computer.

Finally, an operating system is just a big and complicated program built from the same instructions as ordinary programs like Word or a browser. Its task is to control all the other programs that are trying to run, and to manage interactions with the rest of the computer.

This is pretty abstract, so here's a small concrete example of what programming is. Suppose we want to compute the area of a rectangle from its length and width. We might say in English "the area is the product of the length and width." Writing it on the board in school, a teacher could say that area is computed with this formula:

$area = length \times width$

In a higher-level programming language, we would write

```
area = length * width
```

which would be the exact form in most of the popular languages of today. A compiler translates that into a still readable but machine-specific sequence of machine instructions for a computer. That sequence might look like this for a simple hypothetical computer:

```
load       length
multiply   width
store      area
```

Finally a program called an *assembler* converts that sequence of more or less readable instructions into a sequence of machine instructions that can be loaded into the primary memory of a computer; when those instructions are executed, they will compute the area from the given length and width. Of course this glosses over any number of details—how do we specify compiling and loading, how do the length and width get into the computer, how is the area printed, and so on—but it's the essence of the story.

If you prefer to see a working example, here is a complete program in the C programming language that reads a length and width, and prints an area:

```
void main() {
    float length, width, area;
    scanf("%f %f", &length, &width);
    area = length * width;
    printf("area = %f\n", area);
}
```

This program can be compiled and executed on any computer.

Everyone is familiar at least with the names of modern operating systems like Windows and macOS; cell phones run operating systems like Android and iOS.

An operating system is a program that controls a computer, sharing the resources among programs that are running. It manages the primary memory, allocating it to running programs as they need it. On a desktop or laptop the operating system lets you run a browser, a word processor, a music player, and perhaps our little area-computation program, all at the same time, switching its attention to each as necessary.

It also controls the display, giving each program screen visibility when requested, and it manages storage devices like disks, so that when you save your Word document, it is preserved so that you can recover it and resume work later on.

The operating system also coordinates communication with networks like the Internet, so your browser can help you search, check in with friends, shop, and share cat videos, all simultaneously.

It's not so obvious to non-programmers, but an operating system also has to protect programs from other programs in case they have errors, and it has to protect itself from errant or malicious programs and users.

Something similar is going on with the operating systems on phones. Underneath, there's a lot of action to maintain communications via a mobile network or Wi-Fi. Phone apps are exactly the same idea as programs like Word, though often different in detail, and they are written in the same programming languages.

Operating systems today are big and complicated programs. Life was simpler in the 1960s but relative to the time, they were still big and complicated. Typically a computer manufacturer like IBM or DEC (Digital Equipment Corporation) would provide one or more operating systems for its various kinds of hardware. There was no commonality at all between hardware from different manufacturers and sometimes not even between hardware offerings from the same manufacturer, and thus there was also no commonality among operating systems.

To further complicate matters, operating systems were written in assembly language, a human-readable representation of machine instructions, but very detailed and specific to the instruction repertoire of a particular kind of hardware. Each kind of computer had its own assembly language, so the operating systems were big and complicated assembly language programs, each written in the specific language of its own hardware.

This lack of commonality among systems and the use of incompatible low-level languages greatly hindered progress because it required multiple versions of programs: a program written for one operating system had to be in effect rewritten from scratch to move to a different operating system or architecture. As we shall see, Unix provided an operating system that was the same across all kinds of hardware, and eventually it was itself written in a high-level language, not assembly language, so it could be moved from one kind of computer to another with comparatively little effort.

2.2 CTSS and Multics

The most innovative operating system of the time was CTSS, the Compatible Time-Sharing System, which was created at MIT in 1964. Most operating systems in that era were "batch processing." Programmers put their programs on punch cards (this was a long time ago!), handed them to an operator, and then waited for the results to be returned, hours or even days later.

Punch cards were made of stiff high-quality paper and could store up to 80 characters, typically a single line of a program, so the 6-line C program above would require 6 cards, and if a change was necessary, the card(s) would have to be replaced. Figure 2.1 shows a standard 80-column card.

By contrast, CTSS programmers used typewriter-like devices ("terminals" like the Model 33 Teletypes in Figure 3.1 in the next chapter) that were connected directly or by phone lines to a single big computer, an IBM 7094 with

CHAPTER 2: PROTO-UNIX

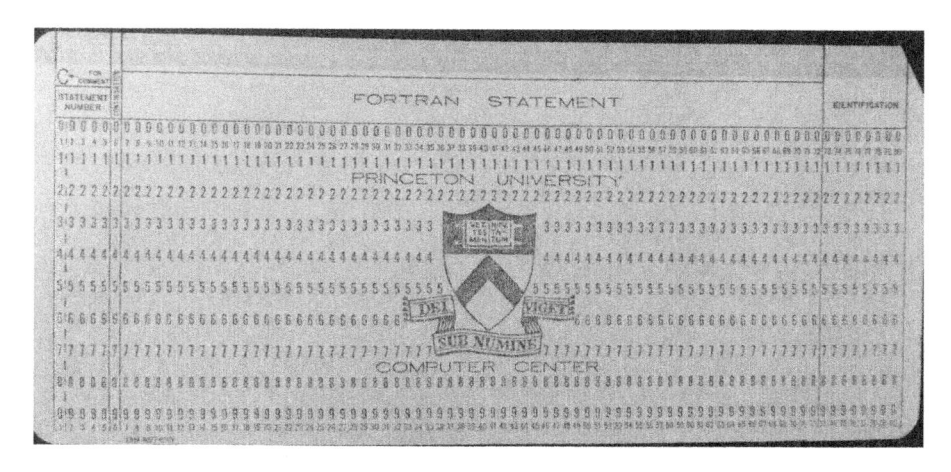

Figure 2.1: Punch card, 7-3/8 by 3-1/4 in [187.325 mm by 82.55 mm]

twice the usual 32K words of memory. The operating system divided its attention among the users who were logged in, switching rapidly from one active user to the next, giving each user the illusion that they had the whole computer at their disposal. This was called "time-sharing," and (speaking from personal experience) it was indescribably more pleasant and productive than batch processing. Most of the time, it really did feel like there were no other users.

CTSS was such a productive programming environment that researchers at MIT decided to create an even better version, one that could serve as an "information utility" to provide computing services to a large and dispersed user population. In 1965, they began to design a system called "Multics," the Multiplexed Information and Computing Service.

Multics was going to be a big job, since it involved ambitious new software, and new hardware with more capabilities than the IBM 7094, so MIT enlisted two other organizations to help. General Electric, which at the time made computers, was to design and build a new computer with new hardware features to better support time-sharing and multiple users. Bell Labs, which had a great deal of experience from creating its own systems since the early 1950s, was to collaborate on the operating system.

Multics was intrinsically a challenging prospect, and it soon ran into problems. In retrospect, it was partly a victim of the *second system effect*: after a success (like CTSS), it's tempting to try to create a new system that fixes all the remaining problems with the original while adding everybody's favorite new features too. The result is often a system that's too complicated, a consequence of taking on too many different things at the same time, and that was the case with Multics. The phrase "over-engineered" appears in several descriptions, and Sam Morgan described it as "an attempt to climb too many trees at once." Furthermore, one does not need to be much of a student of organizations to anticipate that there might also be problems with a project involving two very different companies and a university, in three locations spread across the country.

Half a dozen or more Bell Labs researchers worked on Multics from 1966 through 1969, including Doug McIlroy, Dennis Ritchie, Ken Thompson, and Peter Neumann, who had taken over Vic Vyssotsky's role when Vic moved to another Bell Labs location. Doug was deeply involved with PL/I, the programming language that was to be used for writing Multics software. Dennis had worked on Multics documentation while a student at Harvard and worked on the device input and output subsystem at the Labs. Ken focused on the input/output subsystem, experience that proved valuable when he began work on Unix, though in a 2019 interview, he described his Multics work as being "a notch in a big wheel and it was producing something that I didn't want to use myself."

From the Bell Labs perspective, by 1968 it was clear that although Multics was a good computing environment for the handful of people that it supported, it was not going to achieve its goal of being an information utility that would provide computing services for the Labs at any reasonable cost; it was just too expensive. Accordingly Bell Labs dropped out of the project in April 1969, leaving MIT and GE to soldier on.

Multics was eventually completed, or at least declared a success. It was supported and used until 2000, though not widely. Multics was the source of many really good ideas, but its most lasting contribution was entirely unanticipated: its influence on a tiny operating system called Unix that was created in part as a reaction to the complexity of Multics.

2.3 The origin of Unix

When Bell Labs pulled out of Multics, the people who had been working on it had to find something else to do. Ken Thompson (Figure 2.2) still wanted to work on operating systems, but upper management at the Labs had been burned by the Multics experience and had no interest in buying hardware for another operating system project. So Ken and others spent time exploring ideas and doing paper designs for various operating system components, though with no concrete implementations.

Around this time, Ken found a little-used DEC PDP-7, a computer whose main purpose was as an input device for creating electronic circuit designs.

Figure 2.2: Ken Thompson, ~1984 (Courtesy of Gerard Holzmann)

The PDP-7 was first shipped in 1964 and computers were evolving quickly, so by 1969 it was dated. The machine itself wasn't very powerful, with only 8K 18-bit words of memory (16K bytes), but it had a nice graphics display, so Ken wrote a version of a space-travel game to run on it. A player could wander through the solar system and land on different planets. It was mildly addictive and I spent hours playing with it.

The PDP-7 had another interesting peripheral, a very tall disk drive with a single vertical platter. Credible folklore held that it was potentially dangerous to stand in front of it in case something broke. The disk was too fast for the computer. This presented an interesting problem, and Ken wrote a disk scheduling algorithm that would try to maximize throughput on any disk, but particularly this one.

Now the question was how to test the algorithm. That required loading the disk with data, and Ken decided that he needed a program to put data on it in quantity.

"At some point I realized that I was three weeks from an operating system." He needed to write three programs, one per week: an editor, so he could create code; an assembler, to turn the code into machine language that could run on the PDP-7; and a "kernel overlay –call it an operating system."

Right at that time, Ken's wife went on a three-week vacation to take their one-year-old son to visit Ken's parents in California, and Ken had three weeks to work undisturbed. As he said during a 2019 interview, "One week, one week, one week, and we had Unix." By any measure, this is real software productivity.

A couple of years after Ken and I had both retired from Bell Labs, I asked him about the story that he had written the first version of Unix in three weeks. Here, verbatim, is his email response, which is completely consistent with the much more recent interview:

```
Date: Thu, 9 Jan 2003 13:51:56 -0800

unix was a file system implementation to test thruput and
the like. once implemented, it was hard to get data to it
to load it up. i could put read/write calls in loops, but
anything more sophisticated was near impossible. that was
the state when bonnie went to visit my parents in san diego.

i decided that it was close to a time sharing system, just
lacking an exec call, a shell, an editor, and an assembler.
(no compilers) the exec call was trivial and the other 3
were done in 1-week each - exactly bonnie's stay.

the machine was 8k x 18 bits. 4k was kernel and 4k was
swapped user.

ken
```

This first version of a recognizable Unix system was running in mid to late 1969, so it seems reasonable to say that's when Unix was born.

The early system had a small group of users: Ken and Dennis, of course, plus Doug McIlroy, Bob Morris, Joe Ossanna, and through a bit of blind luck, me. Each user had a numeric user id. Some of these ids corresponded to system functions, not to human users—the root or super-user was id 0, and there were a couple of other special cases. Ids for real users started around 4. I think that Dennis was 5, Ken was 6, and mine was 9. There must be some cachet in having a single-digit user id on the original Unix system.

2.4 What's in a name?

Sometime during these early days the new PDP-7 operating system acquired a name, though the details are murky.

I have a memory of standing in my office doorway, talking with a group that probably included at least Ken, Dennis and Peter Neumann. The system had no name at that point, and (if memory serves) I suggested, based on Latin roots, that since Multics provided "many of everything" and the new system had at most one of anything, it should be called "UNICS," a play on "uni" in place of "multi."

An alternative memory is that Peter Neumann came up with the name UNICS, for "UNiplexed Information and Computing Service." As Peter recalls,

> "I remember vividly that Ken came in one morning for lunch and said that overnight he had written a thousand-line one-user OS kernel for the PDP-7 that Max Matthews had lent him. I suggested that he should make it a multi-user system, and sure enough the next day he came in for lunch having written another thousand lines with a multi-user kernel. It was the one-user kernel that prompted the 'castrated Multics' concept of UNICS."

Peter has graciously said that he doesn't recall further specifics, and so, deservedly or not, I have gotten credit for coining the name.

Either way, UNICS somehow mutated into Unix, which was clearly a much better idea. (It was rumored that AT&T lawyers did not like "Unics" because of its similarity to "eunuchs.") Dennis Ritchie subsequently characterized the name as "a somewhat treacherous pun on Multics," which indeed it is.

2.5 Biography: Ken Thompson

In May 2019, Ken and I had an informal "fireside chat" at the Vintage Computer Festival East in Wall, New Jersey. My role was to ask a few leading questions, then sit back and listen. Some of the material here is paraphrased from that event, which can be found on YouTube.

Ken was born in 1943. His father was in the US Navy, so Ken spent significant parts of his childhood living in different parts of the world, including California, Louisiana, and a few years in Naples.

He grew up interested in electronics, and went to the University of California at Berkeley to study electrical engineering. He said that he found the electronics part really easy because it had been his hobby for 10 years before college. At Berkeley, he discovered the field of computing.

> "I consumed computers, I loved them. At that time, there was no computer science curriculum at Berkeley; it was being invented.
>
> I was drifting along the summer after I graduated. [Graduation] was a surprise because I didn't know that I had gotten all the requirements.
>
> I was just going to stay in the university because ... I owned it. My fingers were in absolutely everything. The main monster computer at the

university shut down at midnight and I'd come in with my key and I'd open it up and it would be my personal computer until 8 AM.

I was happy. No ambition. I was a workaholic, but for no goal."

In his final year, Ken audited a course taught by Elwyn Berlekamp, a Berkeley professor who soon afterwards went to Bell Labs. In the summer after graduating, Ken didn't apply for grad school because he didn't think he was good enough.

"Towards the end of the summer, [Berlekamp] said 'Here's your classes for grad school.' He had applied for me and I got accepted!"

When Ken finished his MS at Berkeley in 1966, Bell Labs was among several companies that tried to recruit him, but he explicitly said that he didn't want to work for a company.

The recruiter kept trying. As Ken says, "I skipped maybe 6 or 8 recruiting attempts by Bell Labs—again, no ambition. The Bell Labs recruiter came and knocked on my door at home. I invited him in; his story is that I offered him ginger snaps and beer." (This must be part of some odd California diet.)

Finally, Ken agreed to come to New Jersey at Bell Labs' expense, but only for a day, and primarily to visit friends from high school days. But when he arrived at Bell Labs, he was impressed by the names he saw:

"The very first thing I did was walk down the corridor of computer science research and every name on the doors on the way down, I knew. It was just shocking. I was interviewed by two amazing people ... one was Shen Lin.

I got in my rent-a-car after the day. Somehow they tracked me, and there was an offer waiting at like the third stop down the east coast. I picked up the offer and drove it from one stop to another, which was maybe two hours, thinking about it, and when I got to the next friend's house, I called them and said OK."

Ken arrived at Bell Labs in 1966 and started work on Multics, then on Unix as described earlier, so I won't repeat that here.

Ken had a long-standing interest in games, and was a chess enthusiast as a kid. He didn't like to lose but at the same time, when he won he felt bad for his opponent, so eventually he became a spectator only. In 1971, he wrote a chess playing program for the PDP-11. This was promising enough that he began to build special-purpose hardware to speed up computation, for example to generate legal moves from a given position. Eventually this culminated in Belle (Figure 2.3), the chess-playing computer that he and Joe Condon evolved from 1976 through 1980.

Figure 2.3: Ken Thompson and Joe Condon (Computer History Museum)

Belle (Figure 2.4) had a successful career. It was the first computer to become a chess master, a rating of 2200 in regular tournament play against human opponents, and it won the 1980 World Computer Chess championship and several ACM computer chess tournaments before retiring to the Smithsonian Institution.

Dennis Ritchie wrote a short article for the International Computer Chess Association on Ken Thompson's activities with various games: *www.bell-labs.com/usr/dmr/www/ken-games.html*. It shows Ken's breadth of game interests, well beyond just chess. There is also a description of Belle's win over Blitz 6.5 at the ACM computer chess championship on December 5, 1978, with comments from Monty Newborn, a computer chess pioneer, and International Master David Levy:

1. e4 e5 2. Nf3 Nc6 3. Nc3 Nf6 4. Bb5 Nd4 5. Bc4 Bc5 6. Nxe5 Qe7

Figure 2.4: Belle chess computer (Courtesy of Computer History Museum)

7. Bxf7+ Kf8 8. Ng6+ hxg6 9. Bc4 Nxe4 10. O-O Rxh2!! 11. Kxh2 {hastening the loss} Qh4+ 12. Kg1 Ng3 13. Qh5 {ineffectual delay} gxh5 14. fxg3+ Nf3# {perhaps uniquely blocking a check, giving a double check and mating simultaneously; "the most beautiful combination created by a computer program to date... computer chess witnessed the start of a new era."}

Chess games end with a win, loss or draw. The 50-move rule says that a player can claim a draw if there have been 50 moves without a capture or a pawn move; this prevents one player from just playing on when there is no way to force a win.

Ken decided to explore the question of whether 50 moves is the right number. He used Belle and some sophisticated database organization to evaluate all four- and five-piece endgames and discovered that some were winnable in more than 50 moves, given optimal play. By this time, Ken was well known in the chess world, and sometimes grandmasters would show up at the Labs to try their hand against Belle, particularly for endgames. I met world champions Anatoly Karpov and Vishy Anand just by being around on the right weekends.

Ken was also an avid pilot, and regularly flew himself and guests around New Jersey, starting from the airport in Morristown. He got other members

CHAPTER 2: PROTO-UNIX 39

of 1127 interested in flying as well, and at peak there were half a dozen private pilots in the "1127 air force." This group used to set off to view fall foliage or fly to some interesting place for lunch. Doug McIlroy recalls:

> "Besides fall foliage in New England, the air force attended an eclipse in the Adirondacks, thanks to Ken's piloting and Rob Pike's telescopes. There was also a flight to observe a transit of Mercury. The astronomical theme in the Unix crew began with Joe Ossanna's `azel`, which had controlled the Telstar ground station, and we used to tell us where to find artificial satellites. Next came Bob Morris's `sky` program, then Ken's celestial event predictor, Lee McMahon's star maps made with my `map` program, and finally Rob's `scat` star catalog."

In December 1992, Ken and Fred Grampp went to Moscow to fly a MiG-29, a step up from their normal Cessnas. Figures 2.5 and 2.6 show Ken ready to go and taxiing back after a flight.

Figure 2.5: Ken Thompson preparing to take off (Courtesy of cat-v.org)

Ken and I retired from Bell Labs at the end of 2000. I went to Princeton, and he joined Entrisphere, a startup founded by Bell Labs colleagues. In 2006, he moved to Google, where with Rob Pike and Robert Griesemer, he created the Go programming language. I heard about his move from

Figure 2.6: Ken taxiing after a flight (Courtesy of cat-v.org)

Entrisphere to Google from someone else, so I asked for confirmation. His reply:

```
Date: Wed, 1 Nov 2006 16:08:31 -0800
Subject: Re: voices from the past

its true. i didnt change the median age of google much,
but i think i really shot the average.
ken
```

Chapter 3

First Edition (1971)

"This manual gives complete descriptions of all the publicly available features of Unix. It provides neither a general overview (see "The Unix Time-sharing System" for that) nor details of the implementation of the system (which remain to be disclosed)."
  First Edition Unix Programmer's Manual, November 3, 1971

"BUGS: rm probably should ask whether a read-only file is really to be removed."
  Section of manual page for rm command, November 3, 1971

The PDP-7 Unix system was interesting enough that people were starting to use it, even though it ran on a tiny computer and didn't have a lot of software. Still, it was clearly useful and had become the preferred computing environment for a small group who found it more fun and productive than the big central computer. Thus Ken Thompson, Dennis Ritchie and others began to lobby for a larger computer that would support more users and enable more interesting research.

One of the early proposals was for the purchase of a DEC PDP-10, which was popular at universities and other research labs. The PDP-10 was loosely similar to the IBM 7090, with 36-bit words like the 7090 and the GE 635 and 645, and it had a great deal more horsepower than the wimpy little PDP-7. But it would also cost much more; the proposal was for half a million dollars.

The Multics experience was an all-too-recent bad memory, so the PDP-10 proposal never got off the ground. As Ken said, the management position was "We don't do operating systems," though perhaps it was more like "We're not going to give you a lot of money for a big machine."

I suppose that it could be argued that one positive role of management is to be cautious most of the time, so that people who want resources are forced to hone their proposals and focus their pitches. If resources are tight, that's more likely to lead to good, well-thought-out work than if there are no constraints.

In any case, the Unix group came up with another idea, to acquire a new minicomputer that DEC had just announced, the PDP-11, which would cost more like $65,000 than $500,000 in 1971 dollars.

This was rejected too. A remark from Sam Morgan in Mike Mahoney's 1989 oral history interview explains some of the reasoning:

> "The management principles here are that you hire bright people and you introduce them to the environment, and you give them general directions as to what sort of thing is wanted, and you give them lots of freedom. Doesn't mean that you always necessarily give them all the money that they want. And then you exercise selective enthusiasm over what they do. And if you mistakenly discourage or fail to respond to something that later on turns out to be good, if it is really a strong idea, it will come back."

In hindsight, being forced to work within constraints was a good thing. As Ken himself said in his 1983 Turing Award lecture,

> "Unix swept into popularity with an industry-wide change from central mainframes to autonomous minis. I suspect that Daniel Bobrow would be here instead of me if he could not afford a PDP-10 and had had to 'settle' for a PDP-11."

(Daniel Bobrow was the primary author of the Tenex operating system, which was written for the PDP-10 in 1969.)

3.1 Unix for patent applications

Direct appeals for a machine failed, but there was an alternative. Bell Labs, as a large and productive scientific research operation, generated a lot of patent applications. At that time, it was being granted an average of almost one patent per day. Patent applications were text documents but with some rigid format requirements, like line-numbered pages. No existing computer system could handle these oddities, so the Patent department was planning to buy hardware from a company which promised software that would eventually produce applications in the proper format, though it could not do numbered lines at the time.

Joe Ossanna came up with another plan. The Patent department would use a PDP-11 for preparing patent applications; the Unix group would write the necessary software for them, complete with a formatting program that would print applications in the proper format; and no, no one would be working on operating systems.

This combination of ideas slalomed around any residual management objections. Money for a PDP-11 came from Max Matthews, the director of the Speech and Acoustics Research Center. Max was supportive because one of his departments heads, Lee McMahon, was very interested in text processing and along with Ossanna was a promoter of the plan.

The deal was approved, a PDP-11 was purchased, and Ken and Dennis quickly converted the PDP-7 version of Unix to run on it. The PDP-11 was a limited machine, with only 24K bytes of primary memory and a half-megabyte disk. The implementation used 16K bytes for the operating system and the remaining 8K for user programs.

Joe Ossanna wrote a program called Nroff ("new roff"), analogous to the existing Roff text formatter, that was able to print patent applications in the required format. By the last half of 1971 typists were cranking out patent applications on Unix full time. Chapter 5 has much more to say about text formatting, since it was a major part of the Unix story in the 1970s.

That was during the day. At night, Ken, Dennis, and others were developing software on the same PDP-11. Development had to be done at night so as not to interfere with the typists, and it had to be careful. The PDP-11 had no hardware protection mechanisms to keep programs from interfering with each other or with the operating system, so a careless mistake could easily crash the system, and an error in the file system could lose everyone's work. But the experience was so successful that the patent department bought another PDP-11 for the Unix group, and that made it possible to do development full time. This version became the first edition of Unix.

Figure 3.1 is a 1972 public-relations picture of Ken Thompson and Dennis Ritchie using a PDP-11 running an early version of Unix. The computer itself is apparently a specific model called a PDP-11/20. The smallish round objects at about head level are DECtapes, which were magnetic tape devices that held 144K 18-bit words. Individual blocks could be read and written, so they could serve as temporary slow but reliable disks; the tapes themselves were removable, so they were used for backup as well.

Ken is typing at a Model 33 Teletype, a sturdy but slow and noisy device, basically a computer-controlled electro-mechanical typewriter that could only print in upper case, at 10 characters per second. The Model 33 dates from 1963 but earlier versions had been in widespread use since the early 1930s.

Teletype Corporation was a part of AT&T, and Teletypes were widely used throughout the Bell System and elsewhere for sending messages, and later for connecting to computers. Whatever was typed on the Teletype keyboard was sent to the computer, and responses were printed (in upper case) on a long roll of paper; the tops of the paper rolls are just visible in the picture. Arguably, one reason why many command names on Unix are short is that it took considerable physical force to type on a Model 33, and printing was slow.

Figure 3.1: Ken (seated) and Dennis at the PDP-11, ~1972 (Wikipedia)

Someone even built an experimental "portable" Model 33. The keyboard and printer were shoehorned into a suitcase-like container that in theory could be carried around, though at 55 pounds (25 kg) you wouldn't carry it far. (It had no wheels either.) It was connected to a remote computer through a dial-up phone connection and a built-in acoustic coupler: plug a telephone handset into a couple of rubber sockets and the coupler converted data into sound and back again, rather like a fax machine. I managed to haul one of these terminals home a couple of times, but calling it portable was too charitable.

Things improved markedly when the Model 37 Teletype came along. It had lower case letters as well as upper case, and was somewhat faster (15 characters per second rather than 10), though it was still strenuous to type on it. It had an extended typebox so it could print mathematical symbols, which was useful for the patent applications as well as our own technical papers, and it could roll the paper back and forth in half-line steps, which made mathematical subscripts and superscripts possible.

Feeding paper into it was also a challenge; it took real contortions to load a new box of fan-fold paper. Bob Morris once sent Joe Ossanna a mail message that consisted of 100 reverse line-feeds; when Joe tried to read the message, the Model 37 sucked the paper back out and deposited it on the floor.

Bob occupied the office across from me for a number of the early years. "Robert Morris" was a common name at Bell Labs; indeed, there was a visitor named Robert Morris in the same office a few years later. So Bob frequently got misdirected mail, which he would dutifully send back, explaining that he was the wrong Morris. One piece of mail kept on coming, an elaborate blueprint from some other part of the company that said "Please initial and return." All attempts to head it off failed, so finally one day Bob did initial it and sent it back. It never appeared again.

3.2 The Unix room

Although each Research MTS had a private office, much Unix development went on in a shared space called "the Unix room." Its actual location changed a couple of times over the years, but it was always a place to hang out, learn what was going on, contribute ideas, or just socialize.

The original Unix room was briefly on the fourth floor of Building 2 where the PDP-7 lived, but the main location for many years was on the sixth floor of Building 2 in room 2C-644. Building 2 only had five office floors. The sixth floor was basically a service corridor: dingy, dimly lit and lined with storage areas holding dusty abandoned equipment in locked wire cages. At one end there was an open area with vending machines that offered appalling coffee and almost inedible pastries that fueled late-night programming, and there were a handful of enclosed spaces, one of which was the Unix room for at least a decade. It held the PDP-11; the picture of Ken and Dennis in Figure 3.1 was taken there. A few tables and chairs and some other terminals made it a good shared working area.

One of the early non-1127 Unix enthusiasts was a very distinguished theoretical physicist, now deceased, who I will call "M— L—." M— L— was eager to use Unix, he was forward-thinking in his use of computers in physics, and he was a kind and generous person. But he would talk your ear off. Once he got started, there was no way to stop him, and you were in for an hour of one-way conversation. So someone scratched a small hole in the frosting of the door to the Unix room so we could peer in before entering, to see if he was there. It was called the "L— hole."

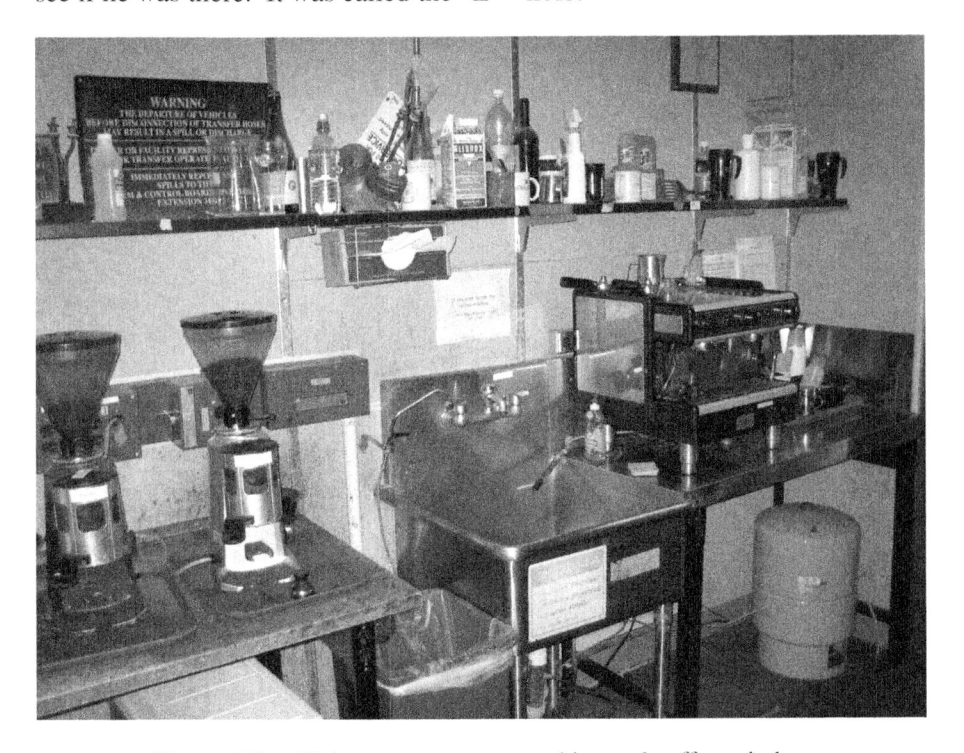

Figure 3.2: Unix room espresso machine and coffee grinders

At some point the Unix room migrated to room 2C-501, on the fifth floor at Stair 9, just around the corner from my office. It also acquired a variety of coffee machines, originally the usual carafes with a heater that kept the coffee warm until it burned (or the carafe did, something that happened regularly), and then a sequence of ever more expensive coffee grinders and espresso machines (Figure 3.2), the last of which cost something like $3,000. If my sources are right, Unix room denizens paid for the machine by passing the hat and management paid for the coffee.

The Unix room was just plain fun; there was always something going on. Some people worked there almost exclusively and rarely used their offices. Others would drop in multiple times a day for coffee and conversation. It's hard to overstate how important the Unix room was for keeping up with what colleagues were doing, and for creating and maintaining a sense of community.

In retrospect, I think that Bell Labs did a good job with space. Private offices, though they cost more than open areas, give people peace and quiet, a place to focus without constant noise in the background, storage for books and papers, and a door to close for intense thought or private conversations. By now, I've spent enough time in open-plan work areas to know that, for me at least, they are utterly destructive of concentration. The Bell Labs mixture of one's own private office and a shared space for the community worked very well.

The Labs also made it easy for people to keep on working at home in the evening. For many years, I had a dedicated phone line (after all, AT&T was the phone company) in my home that let me connect to the Unix systems at Murray Hill so I could work evenings and weekends. As an unexpected fringe benefit, there was a special access code that allowed us to make unlimited and unbilled long distance telephone calls to anywhere in the USA, which was rather a nice perk at a time when long distance calls cost actual money. Ken Thompson told me more on how this came to be:

> "Joe Ossanna decided that we deserved home phone lines and teletypes. He invented a form to order them and made copies that he put in the stationery store room. He put himself down as the approver and then submitted several for the core Unix crowd. After several questioning telephone calls, Joe started getting the forms all filled out, which he approved. It was that simple, he just invented the form and it happened."

In 1985, Peter Weinberger was promoted to department head in 1127, and a professional photograph was taken for the company newspaper, the *Bell Labs News* (familiarly known as the *Bell Labs Good News* since it only printed positive stories). In a serious tactical error, Peter left the original print of his headshot (Figure 3.3) in the Unix room.

Soon his image was all over the place, sometimes filtered through the AT&T logo (Figure 3.4), which had recently been introduced. As Gerard Holzmann says,

> "Within a few weeks after AT&T had revealed the new corporate logo, Tom Duff had made a Peter logo (Figure 3.5) that has since become a symbol for our center. Rob Pike had T-shirts made. Ken Thompson ordered coffee mugs with the Peter logo.

Figure 3.3: Original Peter face (Courtesy of Gerard Holzmann)

Figure 3.4: AT&T "death-star" logo

Over the years, Peter's visage showed up in dozens of places—an organization chart with Peter faces all the way to the top, a large array of circular magnets on stairway walls, impressed in newly laid concrete floors, on microprocessor chips, and, most prominently, in the night of September 16, 1985, on one of the Bell Labs water towers (Figure 3.6).

Rumors have circulated about who the painters were, but lips are still sealed more than three decades later. A reimbursement request for the paint was submitted, but it was rejected. In any case, within a couple of days the water tower image was painted over by an officialdom that apparently did not share our sense of humor.

The full story of Peter's many faces can be found at *spinroot.com/pjw*, a web site maintained by Gerard, who with Rob Pike did many of the original enhancements of Peter's picture.

Figure 3.5: Peter filtered through AT&T logo (Courtesy of Gerard Holzmann)

Bell Labs was an informal place, but sometime in the early to mid 1980s, a new policy was instituted: employees had to wear their badges at all times. This was undoubtedly a sensible precaution to discourage interlopers, but it wasn't popular. As a protest, one colleague, who will remain nameless here, stuck his badge to his forehead with super-glue; another took to wearing it clipped to the hair on his chest, revealing it only upon demand.

The badges had no security features; they consisted only of a picture in a template. Accordingly, we undertook a campaign to create a fake person, Grace R Emlin, who had her own login, `gre`, her own badge (Figure 3.7), and from time to time an appearance on official lists and publications.

I made my own badge with a Mickey Mouse image (Figure 3.8). I wore it regularly, including one day at Bell Labs in Holmdel, New Jersey, for a meeting with Bill Gates, who was there for some marketing of Windows 3.0. No one noticed.

Figures 3.9 and 3.10 show random parts of the Unix room in 2005.

3.3 The Unix Programmer's Manual

One of the early contributions of Unix was its online manual, in a now-familiar format and concise style. Every command, library function, file format, and so on had a page in the manual that described briefly what it was and how to use it. For example, Figure 3.11 shows the 1st Edition manual page for the `cat` command, which concatenates zero or more files onto the standard output stream, by default the user's terminal.

Early man pages tended to be literally only a single page, a slimness that is uncommon today. Besides brevity, a couple of features of man pages were novel at the time, like the BUGS section, which acknowledged that programs

Figure 3.6: Peter on the water tower, 1985 (Courtesy of Gerard Holzmann)

Figure 3.7: Grace Emlin, MTS

do have bugs, or perhaps "features," imperfections that should at least be recorded even if not fixed right away.

The cat command hasn't changed in 50 years, aside from acquiring a few optional and probably unnecessary arguments that somewhat modify its behavior; it's still part of the core set of Unix commands. You can see its current status by typing the command

```
$ man cat
```

in a terminal window in Linux, macOS or Windows Subsystem for Linux (WSL). And of course you can view the manual page for the man command itself with

```
$ man man
```

Figure 3.8: My high-security Bell Labs badge

3.4 A few words about memory

Younger readers might wonder about the accuracy of some of the memory sizes that I have been quoting along the way. For instance, an IBM 7090 or 7094 had 32K (32,768) 36-bit words; the original PDP-7 that Ken used had 8K (8,192) 18-bit words, that is, around one eighth of the 7090 memory; and the first PDP-11 had 24K bytes of primary memory and a half-megabyte disk. For comparison, my 2015 Macbook Air has 8 GB of memory (over 330,000 times as much), a 500 GB disk (half a million times as much), and cost barely a thousand dollars.

In short, memories were tiny by today's standards, where gigabytes of primary memory and terabytes of disk storage are cheap, compact and therefore common. But memory technology in the 1960s and early 1970s was different. The primary memory of a computer was built from arrays of tiny donut-shaped ferrite cores, connected with an intricate though orderly set of wires that were threaded by hand through the cores. Each core could be magnetized one way or the other (think clockwise or counterclockwise) and thus was capable of representing one bit of information; eight cores would be a byte.

Core memory was very expensive, since manufacturing it took highly skilled manual labor, it was bulky, and it weighed a lot. Figure 3.12 shows a

CHAPTER 3: FIRST EDITION

Figure 3.9: Unix room, October 2005

core memory with 16K bits (2K bytes), which in 1971 would have cost about $16,000, or close to a dollar a bit.

Memory was often the most costly component of a computer. When every byte was precious, that scarcity imposed a certain discipline on programmers, who always had to be conscious of how much memory they were using, and sometimes had to resort to trickery and risky programming techniques to fit their programs into the available memory.

One thing that Unix did well was to make effective use of the limited memory of the computers that it ran on. Some of this was due to exceptionally talented programmers like Ken and Dennis, who knew how to save memory.

Some was due to their genius in finding ways to achieve generality and uniformity that made it possible to accomplish more with less code. Sometimes this was achieved by clever programming, while in other cases it was a result of finding better algorithms.

Some was due to the use of assembly language, which at the time did make better use of instructions (run faster) and memory (use less space) than

Figure 3.10: Unix room, October 2005

could be achieved with compilers for high level languages. It was only in the mid 1970s that new memory technology based on semiconductors and integrated circuits became widely available at a price where one could afford the moderate but measurable overhead of high-level languages like C.

Storage allocators like the original `alloc` and Doug McIlroy's later `malloc` library were used to allocate and reallocate memory dynamically as a program was running, another way to make the most of a scarce resource. Naturally this had to be done carefully, since the slightest mistake could cause a program to work incorrectly (something that is not unheard of even today, I might add, at least as I observe students in my classes). Mismanagement of memory remains one of the leading causes of errors in C programs.

When a program failed badly enough, the operating system would notice and would try to help the programmer by producing a file of the contents of main memory—what was in the magnetic cores—from which comes the phrase "core dump," still used though magnetic cores long ago left the scene. The file is still called `core`.

```
         11/3/71                                                   CAT (I)


NAME              cat -- concatenate and print

SYNOPSIS          cat file1 ...

DESCRIPTION       cat reads each file in sequence and writes it on the
                  standard output stream. Thus:

                            cat file

                  is about the easiest way to print a file. Also:

                      cat file1 file2 >file3

                  is about the easiest way to concatenate files.

                  If no input file is given cat reads from the standard input
                  file.

FILES

SEE ALSO          pr, cp

DIAGNOSTICS       none; if a file cannot be found it is ignored.

BUGS

OWNER             ken, dmr
```

Figure 3.11: cat(1) manual page from 1st Edition Unix

3.5 Biography: Dennis Ritchie

This summary of Dennis Ritchie's life is adapted from a memorial that I wrote for the National Academy of Engineering in 2012.

Dennis (Figure 3.13) was born in September 1941. His father, Alistair Ritchie, worked for many years at Bell Labs in Murray Hill. Dennis went to Harvard where he did his undergraduate work in physics and his graduate work in applied mathematics. His PhD thesis topic (1968) was subrecursive hierarchies of functions, and is tough going if one is not an expert, which I am certainly not; Figure 3.14 shows part of one random page, taken from a blurry copy of a draft. Explaining his career path, he said

> "My undergraduate experience convinced me that I was not smart enough to be a physicist, and that computers were quite neat. My graduate school experience convinced me that I was not smart enough to be an expert in the theory of algorithms and also that I liked procedural languages better than functional ones."

As Bjarne Stroustrup, the creator of C++, once said, "If Dennis had decided to spend that decade on esoteric math, Unix would have been stillborn."

Figure 3.12: Magnetic core memory; 16K bits, 2K bytes (~5.25 in, 13 cm)

Dennis had spent several summers at Bell Labs, and joined permanently in 1967 as a member of technical staff in the Computing Science Research Center. For the first few years, he worked on Multics. As noted earlier, Multics proved too ambitious, and as it became clear that it would not live up to its goals, Bell Labs withdrew in 1969, leaving Ken, Dennis and colleagues with experience in innovative operating system design, an appreciation of implementation in high-level languages, and a chance to start over with much more modest goals. The result was the Unix operating system and the C programming language.

The C programming language dates from early in the 1970s. It was based on Dennis's experience with high-level languages for Multics implementation, but much reduced in size because most computers of the time had limited capacity; there simply wasn't enough memory or processing power to support a complicated compiler for a complicated language. This enforced minimality matched Ken and Dennis's preference for simple, uniform mechanisms. C was a good match as well for real computer hardware; it was clear how to translate it into good code that ran efficiently.

Figure 3.13: Dennis Ritchie, ~1984 (Courtesy of Gerard Holzmann)

C made it possible to write the entire operating system in a high level language. By 1973, Unix had been converted from its original assembly language form into C. This made it much easier to maintain and modify the system. It also enabled another giant step, moving the operating system from its original PDP-11 computer to other computers with different architectures. Because most of the system code was written in C, porting the system required not much more than porting the C compiler.

Dennis was a superb technical writer, with a spare elegant style, deft turns of phrase, and often with flashes of dry wit that accurately reflected his personality. He and I wrote *The C Programming Language* together; it was published in 1978, with a second edition in 1988, and has since been translated into more than two dozen languages. Dennis's original C reference manual formed the basis of the ANSI/ISO standard for C that was first produced in 1988, and was a major part of the book. Without doubt, some of the success of C and Unix can be attributed to Dennis's writing.

With Ken Thompson, Dennis received many honors and awards for his work on C and Unix, including the ACM Turing Award (1983), the National Medal of Technology (1999), the Japan Prize for Information and Communications (2011), and the National Inventors Hall of Fame (posthumously in 2019).

Dennis successfully avoided any management role for many years, but finally yielded and became head of the Software Systems department, where he was responsible for the group that was building the Plan 9 operating

Now if $\beta = \omega^m b_m + \cdots + \omega^{n+1} b_{n+1} + \omega^n b_n + \cdots + \omega^0 b_o$, let $\beta' = \omega^m b_m + \cdots + \omega^{n+1} b_{n+1}$. Then $\beta + \omega^n(q + b + 5\underline{m} + 1) = \beta' + \omega^n(q + b + b_n + 5\underline{m} + 1)$ and furthermore β' is the least ordinal with this property. Thus by (3.2),

$$T_P(\bar{x}_k) \leq f_{\beta' + \omega^{n+1}}(q + b + b_n + 5\underline{m} + 1)$$
$$= f_{\beta + \omega^{n+1}}(q + b + b_n + 5\underline{m} + 1) \quad \text{by definition of } \beta'$$
$$\leq f_{\beta + \omega^{n+1}}^{(q+b+b_n+1)}(5\underline{m}) \quad \text{by (3.4.v)}$$
$$\leq f_{\beta + \omega^{n+1}}^{(q+b+b_n+1)} f_1^{(3)}(\underline{m}) \quad \text{by (3.4.ii)}$$
$$\leq f_{\beta + \omega^{n+1}}^{(q+b+b_n+4)}(\underline{m}) \quad \text{by (3.4.vii)}$$

But even if $\underline{m} = 0$, $T_P(\bar{x}_k) = 2 \leq f_{\beta + \omega^{n+1}}^{(q+b+b_n+4)}(\underline{m})$ by (3.4.v). Since $t_{n+1}(\beta) = t_n(\beta) + b_n = b + b_n$, the lemma is proved, concluding Case 4.

Figure 3.14: Excerpt from Dennis Ritchie's PhD thesis (Courtesy of Computer History Museum)

system. Dennis stepped down from management and retired officially in 2007, but continued to come to Bell Labs almost every day until his death in October 2011.

Dennis was modest and generous, always giving credit to others while downplaying his own contributions. A typical example is found in the acknowledgment section of his 1996 retrospective on the evolution of Unix:

> "The reader will not, on the average, go far wrong if he reads each occurrence of 'we' with unclear antecedent as 'Thompson, with some assistance from me.'"

Dennis died in October 2011. This note from his sister and brothers can be found on his home page, which has been preserved at Bell Labs at *www.bell-labs.com/usr/dmr/www.*

> "As Dennis's siblings, Lynn, John, and Bill Ritchie—on behalf of the entire Ritchie family—we wanted to convey to all of you how deeply moved, astonished, and appreciative we are of the loving tributes to

Dennis that we have been reading. We can confirm what we keep hearing again and again:

Dennis was an unfailingly kind, sweet, unassuming, and generous brother—and of course a complete geek. He had a hilariously dry sense of humor, and a keen appreciation for life's absurdities—though his world view was entirely devoid of cynicism or mean-spiritedness.

We are terribly sad to have lost him, but touched beyond words to realize what a mark he made on the world, and how well his gentle personality—beyond his accomplishments—seems to be understood."

Lynn, John, and Bill Ritchie

Chapter 4

Sixth Edition (1975)

"The number of Unix installations has grown to 10, with more expected."
The Unix Programmer's Manual, 2nd Edition, June 1972

"The number of Unix installations is now above 50, and many more are expected."
The Unix Programmer's Manual, 5th Edition, June 1974

The first edition of Unix was up and running by late 1971, if we go by the date on the manual. For the next few years, there was a new edition of the manual roughly every six months, each time with major new features, new tools, and new languages. The 6th edition, whose manual is dated May 1975, was the first that found its way outside of Bell Labs to any significant degree, and it had a major effect on the world.

Unix was first publicly described in a paper by Dennis Ritchie and Ken Thompson, "The Unix Time-Sharing System," that appeared in the Fourth ACM Symposium on Operating Systems Principles in October 1973; it was republished with minor changes in the journal *Communications of the Association for Computing Machinery* (CACM) in July 1974. The abstract begins with a concise summary of a remarkable number of good ideas:

> Unix is a general-purpose, multi-user, interactive operating system for the Digital Equipment Corporation PDP-11/40 and 11/45 computers. It offers a number of features seldom found even in larger operating systems, including:
> (1) a hierarchical file system incorporating demountable volumes;
> (2) compatible file, device, and inter-process I/O;
> (3) the ability to initiate asynchronous processes;
> (4) system command language selectable on a per-user basis; and

(5) over 100 subsystems including a dozen languages.

What were these features "seldom found even in larger operating systems," and what was their significance? The next few sections talk about some of them in more detail. If you're not technically inclined, you can safely skim the chapter; I've tried to summarize the important parts of each section near the beginning so you can skip the details.

4.1 File system

The *file system* is the part of an operating system that manages information on secondary storage like disks, which for many years were sophisticated mechanical devices based on rotating magnetic media, and which today are most often solid state disks and USB flash drives, integrated circuits that have no moving parts.

We are familiar with the abstract view of this information storage through programs like Explorer on Windows and Finder on macOS. Underneath those is a significant amount of software to manage the information on physical hardware, keep track of where each part is, control access, make it efficiently accessible for reading and writing, and ensure that it's always in a consistent state.

Before Multics, most operating systems provided at best complicated and irregular file systems for storing information. The Multics file system was much more general, regular and powerful than other file systems of the time, but it was correspondingly complex. The Unix file system that Ken developed profited from Multics, but was significantly simpler. Its clean, elegant design has over the years become widely used and emulated.

A Unix file is simply a sequence of bytes. Any structure or organization of the contents of a file is determined only by the programs that process it; the file system itself doesn't care what's in a file. That means that any program can read or write any file. This idea seems obvious in retrospect, but it was not always appreciated in earlier systems, which sometimes imposed arbitrary restrictions on the format of information in files and how it could be processed by programs. Doug McIlroy describes one example:

> "Typically source code was a distinguished type, different from data. Compilers could read source, compiled programs could read and write 'data.' Thus the creation and inspection of Fortran programs was often walled off from the creation and inspection of other files, with completely different ways to edit and print them. This ruled out the use of programs to generate (or even simply copy) Fortran programs."

Unix made no such distinctions: any program could process any file. If applying a program to a file doesn't make sense—for example, trying to compile a Fortran source file as if it were C—that doesn't have anything to do with the operating system.

Unix files are organized in directories. (Other operating systems often call these folders.) A Unix directory is also a file in the file system, but one whose contents are maintained by the system itself, not by user programs. A directory contains information about files, which may in turn be directories.

A Unix directory entry includes the file name within the directory, access permissions, file size, date and time of creation and modification, and information about where to find the contents of the file. Each directory has two special entries named "." (the directory itself) and "..", the parent directory; these are pronounced "dot" and "dotdot." The root directory is the top of this hierarchy; its name is /. Any file can be reached by following the path down from the root, and the root can be found from any file by going up the sequence of .. parent directories. Thus the text for this book might be found in /usr/bwk/book/book.txt. The system also supports the notion of a current directory, so that filenames can be relative to the current location in the file system, rather than having to specify a full path from the root.

Because any directory can contain subdirectories, the file system can be arbitrarily deep. This organization of nested directories and files is called a "hierarchical" file system. Again, though the advantages are obvious in hindsight, hierarchical file systems were not widely available before Multics and then Unix. For example, some file systems limited the depth of nesting; CTSS limited it to two levels.

4.2 System calls

An operating system provides a set of services to programs that run on it, services like starting and stopping programs, reading or writing information in files, accessing devices and network connections, reporting information like date and time, and many others. These services are implemented within the operating system, and are accessible from running programs through a mechanism called *system calls*.

In a very real sense, the system calls *are* the operating system, since they define what services the system provides. There may be multiple independent implementations of a set of system calls, as is the case with different versions of Unix and Unix-like systems. A completely different operating system, say Windows, could provide software to convert Unix system calls into its own system calls. And there are sure to be system calls that are unique to a particular operating system even if it is Unix-like.

The first edition of Unix had just over 30 system calls, about half of which were related to the file system. Because files contained only uninterpreted bytes, the basic file system interface was dead simple, only five system calls to open or create a file, read or write its bytes, and close it. These were accessed from a C program by calling functions using statements like these:

```
fd = creat(filename, perms)
fd = open(filename, mode)
nread = read(fd, buf, n)
nwrite = write(fd, buf, n)
status = close(fd)
```

The `creat` system call creates a new file and sets its access permissions, which normally would include or exclude the ability to read, write and execute for the user, for the user's group, and for everyone else. These nine bits give considerable control with comparatively little mechanism. The `open` system call opens an existing file; `mode` specifies whether the file is to be read from or written to, and `filename` is an arbitrary path in the hierarchical file system.

The `fd` value that results from calling `open` and `creat` is called a *file descriptor*, a small non-negative integer that is used in subsequent reads and writes of the file. The `read` and `write` system calls attempt to transfer a specified number `n` of bytes either from the file or to the file; each function returns the number of bytes that were actually transferred. For all of these system calls, if a negative value (usually −1) is returned, that indicates some kind of error.

By the way, the `creat` system call really is spelled that way, for no good reason other than Ken Thompson's personal taste. Rob Pike once asked Ken what he would change if he were to do Unix over again. His answer? "I'd spell `creat` with an e."

Another Unix innovation was to have peripheral devices like disks, terminals, and others appear as files in the file system; disks are the "demountable volumes" mentioned in the list of features. The same system calls are used to access devices as are used to access files, so the same code can manipulate files and devices. Of course it isn't that simple, since real devices have weird properties that must be handled, so there are other system calls for dealing with the idiosyncrasies, especially of terminals. This part of the system was not pretty.

There are also system calls for setting a position within a file, determining its status, and the like. These have all been embellished and occasionally improved over 50 years, but the basic model is simple and easy to use.

It may be hard for today's readers to appreciate just how much of a simplification all of this was. In early operating systems, all of the myriad complexities of real devices were reflected through to users. One had to know all about disk names, their physical structure like how many tracks and cylinders they had, and how data was organized on them. Steve Johnson reminded me of how awkward it was with the time-sharing subsystem on the main Honeywell computer at this time:

> "The Honeywell TSS system required you to enter a subsystem to create a disk file. You were asked about 8 questions: initial size of file, maximum size, name, device, who could read it, who could write it, etc. Each of these had to be answered interactively. When all the questions had been answered, the operating system was given the information, and, likely as not, something was mistyped and the file creation failed. That meant you got to enter the subsystem again and answer all the questions again. It's no wonder that when a file finally got created, the system said 'SUCCESSFUL!'"

Unix followed the example of Multics in hiding all of this irrelevant nonsense: files were just bytes. The user determined what the bytes were, while the operating system looked after storing and retrieving them, without exposing device properties to users.

4.3 Shell

The shell is a program that runs other programs. It's the program that lets users run commands, and is the primary interface between users and the operating system. When I log in to a Unix system, my keyboard is connected to a running instance of the shell. I can type commands, usually one at a time, and the shell runs each in turn, and after each one completes, it's ready for the next. So a session might look like this, where $ is a prompt that the shell prints so I know it's waiting for me to do something. What I type is in *this slanted font*.

```
$ date                    tell me the date and time
Tue Dec  3 07:38:24 EST 2019
$ ls                      list the contents of the directory
book.pdf
book.txt
$ wc book.txt             count the lines, words and characters in book.txt
    9920    59412   362867 book.txt
$ cp book.txt backup.txt  copy book.txt to a backup file
```

One very important note: the shell is an ordinary user program, not some integral part of the operating system, another idea taken from Multics. (That's the "system command language selectable on a per-user basis" mentioned in the feature list.) Because the shell is a user program, it's easily replaced by a different one, which is why there are so many Unix shells. If you don't like the way one shell works, you can pick another or even write your own and use it instead. Speaking of "the shell" is not specific.

That said, all Unix shells provide the same basic features, usually with the same syntax. The most important feature is to run programs. They all also provide filename *wildcards*, where pattern metacharacters like "`*`" are expanded into a list of filenames that match the pattern. For instance, to run the program `wc` (word count) to count the lines, words and characters in all the files in the current directory whose names begin with `book`, one gives the command

```
$ wc book*
```

The shell expands the pattern `book*` into all the filenames in the current directory that match any name beginning with `book`, and runs `wc` with those names as arguments. The `wc` command itself doesn't know that the list of filenames was specified by a pattern. It's important that expansion is done by the shell, not by individual programs. For many years, Microsoft's MS-DOS operating system didn't work that way, so some programs did their own expansions and others didn't; users couldn't count on consistent behavior.

Another major service of the shell is input/output redirection. If a program reads from the standard input (by default the terminal), it can be made to read from an input file instead by

```
$ program <infile
```

and if it writes to the standard output (again by default the terminal), its output can be directed into an output file with

```
$ program >outfile
```

The output file is created if it doesn't already exist. As with filename expansion, the program doesn't know that its input or output is being redirected. This is a uniform mechanism, applied by the shell, not by individual programs, and easier to use than approaches like specifying file input and output by parameters for filenames, as in

```
$ program in=infile out=outfile
```

A *shell script* is a sequence of commands that have been stored in a file. Running an instance of the shell with this file as input runs the commands of the script as if they had been typed directly:

    $ sh <scriptfile

A script encapsulates a command sequence. For example, for this book, I run a sequence of simple checks that look for spelling and punctuation errors, improper formatting commands, and other potential gaffes. Each of these checks runs a program. I could type the commands over and over again, exactly the same each time. But instead, I can put the sequence of commands in a single script file called check and thus can check the book by running a single command. Other scripts print the book and make a backup copy.

These scripts are in effect new Unix commands, though highly specialized to me and this particular book. Such personal commands are a common use of shell scripts, a way to create shorthands for one's own frequent computations. I still use some scripts that I wrote 30 or 40 years ago, and this is not at all unusual among long-time Unix users.

The final step in making shell programs fully equivalent to compiled programs was to make it so that if a file was marked executable, it would be passed to a shell for execution. In this way, shell scripts became first-class citizens, indistinguishable from compiled programs when they were executed:

    $ check book.txt

Shell scripts do not replace compiled programs, but they are an important part of a programmer's toolkit, both for personal use and for larger tasks. If you find yourself doing the same sequence of commands over and over again, then you put them into a shell script and thus automate away some drudgery. If a shell script proves to be too slow, it can be rewritten in another language. We'll see more of the power of shell scripts as we examine pipes in the next section.

4.4 Pipes

Pipes are perhaps the single most striking invention in Unix. A pipe is a mechanism, provided by the operating system and made easily accessible through the shell, that connects the output of one program to the input of another. The operating system makes it work; the shell notation to use it is simple and natural; the effect is a new way of thinking about how to design and use programs.

The idea of connecting programs has been around for a long while. One of the clearest statements in the Unix context appeared in an internal document that Doug McIlroy wrote in 1964, advocating, among other things, the idea of screwing programs together "like garden hose." Figure 4.1 is from a well-worn page that hung on the wall in my office at Bell Labs for 30 years. By the way, notice the typing errors and the generally terrible quality. This illustrates what typewritten documents often looked like. The bottom part of the figure is a corrected transcription:

Summary--what's most important.

To put my strongest concerns in a nutshell:

1. We should have some ways of ccupling programs like
garden hose--screw in another segment when it becomes then.
it becomes necessary to massage data in another way.

```
            Summary--what's most important
        To put my strongest concerns in a nutshell:
    1.  We should have some ways of coupling programs like
    garden hose--screw in another segment when it becomes
    necessary to massage data in another way.
```

Figure 4.1: Doug McIlroy's pipe idea (1964)

Doug wanted to allow arbitrary connections in a sort of mesh of programs, but it was not obvious how to describe an unconstrained graph in a natural way, and there were semantic problems as well: data that flowed between programs would have to be queued properly, and queues could explode in an anarchic connection of programs. And Ken couldn't think of any real applications anyway.

But Doug continued to nag and Ken continued to think. As Ken says, "One day I got this idea: pipes, essentially as they are today." He added a pipe system call to the operating system in an hour; he describes it as "super trivial" given that the mechanisms for I/O redirection were already there.

Ken then added the pipe mechanism to the shell, tried it out, and called the result "mind-blowing."

The pipe notation was simple and elegant, a single character (the vertical bar `|`) between a pair of commands. So, for example, to count the number of files in a directory, pipe the output of `ls` (one line per file) into the input of `wc` (counts the lines):

```
$ ls | wc
```

It is often appropriate to think of a program as a *filter* that will read data in, process it in some way, and write the output. Sometimes this is perfectly natural, as in programs that select or mutate or count things on the fly, but sometimes the filter does not operate on the fly. For example, the `sort` command necessarily has to read all its input before it can produce any output, but that's irrelevant—it still makes sense to package it as a filter that can fit into a pipeline.

Ken and Dennis upgraded every command on the system in a single night. The major change was to read data from the standard input stream when there were no input filename arguments. It was also necessary to invent stderr, the standard error stream. The standard error was a separate output stream: error messages sent to it were separated from the standard output and thus would not be sent down a pipeline. Overall, the job was not hard—most programs required nothing more than eliminating extraneous messages that would have cluttered a pipeline, and sending error reports to stderr.

The addition of pipes led to a frenzy of invention that I remember vividly. I don't have an exact date but it would have been in the second half of 1972, since pipes were not in the second edition of the manual (June 1972) but were present in the third edition (February 1973).

Everyone in the Unix room had a bright idea for combining programs to do some task with existing programs rather than by writing a new program. One of mine was based on the `who` command, which lists the currently logged-in users. A command like `who` isn't terribly relevant today when most people work on their own computer, but since the essence of time-sharing was *sharing* the same computer, it was helpful to know who else was using the system at the same time. Indeed, `who` added to the sense of community: you could see who was also working, and perhaps get help if you had a problem, even if both parties were at their homes late at night.

The `who` command prints one line for each logged-in user, `grep` finds all occurrences of a specific pattern, and `wc` counts the number of lines, so these pipelines report the state of logged-in users.

```
who                      # who is logged in?
who | wc                 # how many are logged in?
who | grep joe           # is joe logged in?
who | grep joe | wc      # how many times is joe logged in?
```

To see what an improvement pipes were, consider how the last task would be performed without pipes, using input and output redirection to files:

```
who >temp1
grep joe <temp1 >temp2
wc <temp2
```

followed by removing the temporary files. Pipes make this into a single command without temporary files.

Ken's favorite pipe example was a speaking desk calculator that used Bob Morris's dc calculator program. Ken's program number printed numbers as words ("127" became "one hundred and twenty seven"), and speak synthesized speech from its input. As Ken said in an interview in 2019,

"You typed *1 2 +* into dc, which was piped to number, which was piped to speak, and it said 'four.'"

[Laughter]

"I was never good at math."

Pipes are one of the foremost of Unix contributions, and obvious only in retrospect. As Dennis said in "The Evolution of the Unix Time-sharing System" in 1984,

"The genius of the Unix pipeline is precisely that it is constructed from the very same commands used constantly in simplex fashion. The mental leap needed to see this possibility and to invent the notation is large indeed."

4.5 Grep

Unix began life as a commandline system, that is, one where users type commands to run programs, rather than pointing at and clicking icons with a mouse, as is normal when using Windows or macOS. A commandline interface is not as easy as point and click for novices, but in the hands of someone with even moderate experience, it can be far more effective. It allows for automation that's not possible with a graphical interface: sequences of commands can be run from scripts and applied to large numbers of files with a single command.

Unix has always had a rich collection of small commandline tools, that is, programs that handle simple frequently occurring tasks. Half a dozen commands manipulate the file system, like ls for listing the files in a directory, rather like Finder on a Mac or Explorer on Windows; cat and cp for copying files in various ways; mv ("move") for renaming; and rm for removing files. There are commands for processing file contents, like wc for counting things, `sort` for sorting files, plus a handful for comparing files, another few for transformations like case conversion, and some for selecting parts of files. (Unix users will recognize `uniq`, `cmp`, `diff`, `od`, `dd`, `tail`, `tr`, and `comm`.) Add another dozen tools that don't fit into those categories, and you have a repertoire of 20 or 30 commands that let you do all kinds of basic tasks easily.

In effect, the tools are verbs in a language, and files are the nouns that the verbs apply to. The language is often irregular, and each command has optional arguments that modify its behavior; for example, `sort` normally sorts by line in alphabetic order, but arguments can change that so it sorts in reverse order, numerically, on specific fields, and so on.

To use Unix well, one must learn what amounts to a family of irregular verbs, just as for natural languages. Naturally there are frequent complaints about the historical irregularities, but occasional attempts to rationalize them have for the most part not been very successful.

The prototypical tool, the command that started us thinking about "tools" more than just "programs," was grep, the pattern-searching program originally written by Ken Thompson. As Ken said about grep in 2019,

> "I had it written but I didn't put it in the central repository of programs because I didn't want people to think that I dictated what was there.
>
> Doug McIlroy said 'Wouldn't it be great if we could look for things in files?' I said 'Let me think about it overnight' and the next morning I showed him the program that I had written earlier. He said 'That's exactly what I wanted.'
>
> Since then grep has become a noun and a verb; it's even in the OED. The hardest part was naming it; it was originally 's', for search."

The name grep comes from a command in the `ed` text editor, `g/re/p`, that prints all lines that match the regular expression pattern `re`; the Oxford English Dictionary entry for grep (Figure 4.2) has it right. (Since the OED has blessed grep as a legitimate English word, I'm going to use it without special fonts or capitalization.)

My personal favorite grep story comes from a day in 1972 when I was called by someone at the Labs who said

> "I noticed that when I hold my new pocket calculator upside down, some of the numbers make letters; for example, 3 becomes E and 7 becomes L. I know you guys have a dictionary on your computer. Is there any way you can tell me what words I can make on my calculator

grep, n.

Text size: A A

View as: Outline | Full entry

Quotations: Show all | Hide all Keywords: On | Off

Pronunciation: Brit. ▶/grɛp/, U.S. ▶/grɛp/
Frequency (in current use): ● ● ● ●
Origin: Formed within English, by conversion. **Etymon:** English *grep*.
Etymology: < *grep*, a string of characters used as a command in the Unix operating system < the initial letters of *g*lobal(ly) search *r*egular *e*xpression *p*rint.
The string *g/re/p* (where *re* stands for the regular expression searched for) was earlier used in a Unix text editor as the syntax for a sequence of commands performing the same operation as *grep*.

(Show Less)

Computing.

A Unix command used to search files for the occurrence of a string of characters that matches a specified sequence or pattern, and to output all the lines matching this. Also **grep command**.

Categories »

> 1973 *V4/man/man1/grep.1* in *Unix Version 4* (Electronic text)   *Grep* will search the input file (standard input default) for each line containing the regular expression. Normally, each line found is printed.

Figure 4.2:   OED entry for grep

when I hold it upside down?"

Figure 4.3 shows what he had in mind.

Being in a research group, it felt good to be able to help someone who had a really practical problem. So I asked him what letters he could make when he held his calculator upside down, and he said "BEHILOS". I turned to my keyboard and typed this command:

```
grep '^[behilos]*$' /usr/dict/web2
```

The file /usr/dict/web2 is the word list from Webster's Second International dictionary—234,936 words, one per line—and the cryptic string of characters between quotes is a *regular expression* or pattern that in this case specifies lines that contain only arbitrary combinations of any of these seven letters and nothing else.

Out came the most amazing list, the 263 words in Figure 4.4. I'm a native speaker of English, but that list has a fair number of words I've never seen

Figure 4.3: BEHILOS on a calculator

before. In any case, I printed them out and sent them to the guy. I think he must have been satisfied; I never heard from him again. But he left me with a great story and a wonderful demonstration of the value of tools like grep and notations like regular expressions.

Over time, the word grep became a noun, a verb, a gerund (grepping), and part of everyday speech in the Unix community. Have you ever grepped your apartment for your car keys? Bumper stickers and t-shirts riffed on the AT&T commercial that invited people to "reach out and touch someone":

"Reach out and grep someone."

Arno Penzias, Nobel Prize winner and, as vice president of Research, my boss three levels up, called me one day to ask whether it was safe for him to use this phrase in a public talk.

4.6 Regular expressions

I've used the phrase "regular expression" without really explaining it. A regular expression is a notation for specifying a pattern of text. It could be as simple as a word like *expression* or a phrase like *regular expression* or something significantly more complicated. In effect, a regular expression is a small language for describing text patterns. In the usual notations, a word or phrase is a regular expression that stands for itself in text, and a regular

b	be	bebless	beboss	bee	beeish	beelol	
bees	bel	belee	belibel	belie	bell	belle	
bes	besee	beshell	besoil	bib	bibb	bibble	
bibi	bibless	bilbie	bilbo	bile	bilio	bill	
bilo	bilobe	bilsh	bios	biose	biosis	bis	
bleb	blee	bleo	bless	blibe	bliss	blissless	
blo	blob	bo	bob	bobbish	bobble	bobo	
boho	boil	bole	bolis	boll	bolo	boo	
boob	boohoo	bool	boose	bose	bosh	boss	
e	ebb	eboe	eel	eelbob	eh	el	
elb	ell	elle	els	else	es	ess	
h	he	heel	heelless	hei	heii	helbeh	
hele	helio	heliosis	hell	hellhole	hellish	hello	
heloe	helosis	hi	hie	hill	his	hish	
hiss	ho	hob	hobbil	hobble	hobo	hoe	
hoi	hoise	hole	holeless	holl	hollo	hoose	
hoosh	hose	hosel	hoseless	i	ibis	ibisbill	
ie	ihi	ill	illess	illish	io	is	
isle	isleless	iso	isohel	issei	l	lee	
lees	lei	less	lessee	li	libel	libelee	
lie	liesh	lile	lill	lis	lish	lisle	
liss	lo	lob	lobbish	lobe	lobeless	lobo	
lobose	loess	loll	loo	loose	loosish	lose	
losel	losh	loss	lossless	o	obe	obese	
obi	oboe	obol	obole	obsess	oe	oes	
oh	ohelo	oho	oii	oil	oilhole	oilless	
oleo	oleose	olio	os	ose	osse	s	
se	see	seel	seesee	seise	sele	sell	
sellie	sess	sessile	sh	she	shee	shell	
shi	shiel	shies	shih	shill	shilloo	shish	
sho	shoe	shoebill	shoeless	shole	shoo	shooi	
shool	si	sib	sie	sil	sile	sill	
silo	siol	sis	sise	sisel	sish	sisi	
siss	sissoo	slee	slish	slob	sloe	sloo	
sloosh	slosh	so	sob	soboles	soe	soh	
soho	soil	soilless	sol	sole	soleil	soleless	
soles	soli	solio	solo	sool	sooloos	sosh	
soso	sosoish	soss	sossle				

Figure 4.4: Make these on your calculator upside-down

expression recognizer will find that word wherever it occurs.

Regular expressions also make it possible to specify more complicated patterns by giving special meanings to some characters, called metacharacters. For example, in grep, the metacharacter "." will match any single character, and the metacharacter "*" will match any number of repetitions of the preceding character, so the pattern (.*) will match any sequence of characters enclosed in parentheses.

Unix has had a long love affair with regular expressions, which are pervasive in text editors, in grep and its derivatives, and in many other languages and tools. Regular expressions of a slightly different flavor are also used in filename patterns as seen in shell "wildcards" that match groups of filenames.

Ken Thompson's QED editor on Multics and later the GE 635 (where I first encountered it) used regular expressions, and Ken invented a very fast algorithm that allowed it to process arbitrarily complicated expressions quickly. The algorithm was even patented. QED was sufficiently powerful that in principle it was possible to write any program just using editor commands (though no one in their right mind would do so). I even wrote a tutorial on QED programming; that was largely wasted effort, but it started me writing such documents.

QED was overkill for most purposes. The Unix ed text editor that Ken and Dennis wrote originally and several other people subsequently modified (even me) was much simpler than QED, but it too had regular expressions. Since grep was derived from ed, its regular expressions were the same as those in ed.

A variant style of regular expressions is used for filename wildcards. Although wildcards are interpreted by the shell, because primary memory on the PDP-7 was so limited, the first implementation was a separate program called glob (for "global") called by the shell, and the act of generating an expanded list of filenames from a pattern was called "globbing." The name glob lives on in libraries in several programming languages today, including Python.

One of Al Aho's early contributions to Unix was an extension of grep that allowed a richer class of regular expressions, for example making it possible to search for alternatives like this|that. Al called the program *egrep*, for "extended grep."

It's worth saying a bit more about egrep, since the story behind it is an excellent example of the kind of interplay between theory and practice, and of typical interactions among members of 1127, that led to so much good software. This story comes from Doug McIlroy:

"Al Aho's first egrep was a routine implementation of an algorithm from *The Design and Analysis of Computer Algorithms* by Aho, Hopcroft and Ullman. I promptly used it for a calendar program that used a huge automatically generated regular expression to recognize date patterns like 'today,' 'tomorrow,' 'until the next business day,' and so on, represented in a large range of date styles.

To Al's chagrin, it took something like 30 seconds to compile into a recognizer that would then run in no time at all.

He then came up with the brilliant strategy of generating the recognizer lazily as parts of it were needed, rather than all in advance, so only a tiny fraction of the exponentially vast number of states were ever constructed. That made a tremendous difference: in practice, egrep always ran fast no matter how complex the pattern it was dealing with. So egrep is distinguished by technical brilliance, invisible unless you know how badly standard methods can perform."

This is a common Unix story: a real problem from a real user, deep knowledge of relevant theory, effective engineering to make the theory work well in practice, and continuous improvement. It all came together because of broad expertise in the group, an open environment, and a culture of experimenting with new ideas.

4.7 The C programming language

New programming languages have been a big part of Unix since the very beginning.

One of the most important contributions of Multics was its attempt to write the operating system in a high-level language, PL/I. PL/I was created in 1964 by IBM to try to combine all the good ideas from Fortran, Cobol and Algol into a single language. It turned out to be a grand example of the second system effect. The language was too big and complicated for most programmers, it was hard to compile, and a working compiler for Multics was not delivered on time. As an interim measure, Doug McIlroy and Doug Eastwood created a simplified subset called EPL ("Early PL/I") for use with Multics, but it was still a complicated language.

BCPL ("Basic Combined Programming Language") was another language intended for system programming. It had been designed by Martin Richards, a professor at the University of Cambridge, and he wrote a compiler for it while visiting MIT in 1967. BCPL was much simpler than any dialect of PL/I, and was well suited for writing operating system code. Members of the Bell Labs Multics effort were very familiar with BCPL.

When the Labs pulled out of Multics, Ken Thompson decided that "no computer is complete without Fortran," so he started to write a Fortran compiler for the PDP-7. This proved too tough, since PDP-7 Unix had only 4K 18-bit words (8 KB) of primary memory for user programs like a compiler.

Ken kept redesigning, eventually coming up with a language that did fit the PDP-7 and was much closer to BCPL than to Fortran. He called it B. As Dennis Ritchie explained in "The Development of the C Language" in 1993,

> "B can be thought of as C without types; more accurately, it is BCPL squeezed into 8K bytes of memory and filtered through Thompson's brain. Its name most probably represents a contraction of BCPL, though an alternate theory holds that it derives from Bon, an unrelated language created by Thompson during the Multics days. Bon in turn was named either after his wife Bonnie, or (according to an encyclopedia quotation in its manual), after a religion whose rituals involve the murmuring of magic formulas."

Up to this point, the computers in our story have been "word-oriented," not "byte-oriented." That is, they manipulate information in chunks that are significantly larger than a single byte. The IBM 7090 and similar computers like the GE series manipulated information naturally only in chunks that were 36 bits (roughly four bytes); PDP-7 chunks were 18 bits (two bytes). Word-oriented computers were clumsy for processing bytes individually or in sequences: programmers had to use library functions or go through programming contortions to access the individual bytes that were packed into the larger chunks.

By contrast, the PDP-11 was byte-oriented: its fundamental unit of primary memory was the 8-bit byte, not the 18- or 36-bit words of earlier computers, though it could also manipulate information in larger chunks like 16- and 32-bit integers and 16-bit addresses.

B was a good fit for word-oriented computers like the PDP-7 but not for byte-oriented ones like the PDP-11, so when the PDP-11 arrived, Dennis started to enhance B for the new architecture and to write a compiler for it. The new language was called "NB," for "New B," and eventually it became C.

One of the main differences was that where B was typeless, C supported data types that matched the data types that the PDP-11 provided: single bytes, two-byte integers, and eventually floating-point numbers with four or eight bytes. And where languages like BCPL and B treated pointers (memory addresses) and integers as the same, C formally separated them, though for many years programmers unwisely treated them as if they were the same size.

One of C's novel contributions to programming languages was the way that it supported arithmetic operations on typed pointers. A pointer is a value corresponding to an address, that is, a location in primary memory, and it has a type, the type of the objects that it will point to. If that location corresponds to an element of an array of that particular type of object, then in C, adding 1 to the pointer yields the address of the next element of the array. Although careless use of pointers is a recipe for broken code, pointer arithmetic is natural and works well when used correctly.

It had been clear for some time that Unix should be converted from assembly language to a higher level language, and C was the obvious choice. Ken tried three times in 1973 to write the kernel in C but it proved too difficult until Dennis added a mechanism for defining and processing nested data structures (`struct`) to the language. At that point, C was sufficiently expressive for writing operating system code, and Unix became mostly a C program. The 6th edition kernel has about 9,000 lines of C and about 700 lines of assembly language for machine-specific operations like setting up registers, devices and memory mapping.

The first widely distributed description of C was *The C Programming Language* (Figure 4.5), a book that Dennis and I published in 1978; a second edition was published in 1988.

I had learned B rather superficially, and for my own amusement wrote a tutorial to help others to learn it too. When Dennis created C, it was easy to modify the B tutorial to make one for C. The C tutorial proved to be popular, and as Unix and C spread, I thought that it would be worth trying to write a book about C. Naturally I asked Dennis if he would write it with me. He might have initially been reluctant, but I twisted his arm harder and eventually he agreed. Getting Dennis to work on the book was the smartest or maybe just the luckiest thing I ever did in my technical career—it made the book authoritative because Dennis was a co-author, and it allowed me to include his reference manual.

I wrote the first drafts of most of the tutorial material originally, but Dennis wrote the chapter on system calls, and of course provided the reference manual. We made many alternating passes over the main text, so that's a blend of our styles, but the reference manual stayed almost exactly as it had been, a pure example of Dennis's writing. It describes the language with what Bill Plauger once called "spine-tingling precision." The reference manual is like C itself: precise, elegant, and compact.

The first official C standard, from ANSI, the American National Standards Institute (and also from ISO, the International Standards Organization) was completed in 1989. Its description of the language was directly based on Dennis's reference manual. Dennis was involved in the early stages of the first C standard, where his standing as the creator of the language gave his opinions weight, and he was able to head off one or two really bad proposals.

CHAPTER 4: SIXTH EDITION

Figure 4.5: Cover of K&R first edition, 1978

The C language is important, but so also is its use of a standard library that provides the basic facilities that programmers need for operations like formatted input and output, string processing, and mathematical functions. C came with a modest-sized library of such functions, so that programmers did not need to reinvent each routine when they wrote a new program.

The largest library component provided formatted output, familiar today to all programmers through C's `printf` function, which has been adopted into many other languages. Mike Lesk's portable I/O package, written in 1972 so that programs could be easily moved to and from Unix, contained the first version of `printf` and also included `scanf` for parsing formatted input. These were reworked and included with the C compiler.

Although `printf` and `scanf` have had extensions since, the core set of conversions work as they did in the early 1970s, as do most of the other functions in the library. Today, the standard library is just as much a part of the C standard as is the language specification itself.

It's interesting to contrast the C approach to other languages. In Fortran and Pascal, for example, input and output are part of the language, with special syntax for reading and writing data. Some other languages don't include input and output and at the same time don't provide a standard library, which is probably the least satisfactory choice.

C has been very successful, one of the mostly widely used languages of all time. Although it began life on PDP-11 Unix, it has spread to essentially every kind of computer that exists. As Dennis said in a paper for the second *History of Programming Languages* conference in 1993,

> "C is quirky, flawed, and an enormous success. While accidents of history surely helped, it evidently satisfied a need for a system implementation language efficient enough to displace assembly language, yet sufficiently abstract and fluent to describe algorithms and interactions in a wide variety of environments."

Of course there are many programming languages, often with noisy adherents and detractors, and C comes in for its share of criticism. It remains the core language of computing, however, and is almost always in the top two or three in lists of language popularity, influence and importance. To my mind, no other language has ever achieved the same balance of elegance, expressiveness, efficiency and simplicity. C has also inspired the basic syntax of many other languages, including C++, Java, JavaScript, Awk and Go. It has been an exceptionally influential contribution.

4.8 Software Tools and Ratfor

By mid to late 1975, Unix had been publicly described at conferences and in journal papers and the 6th edition was in use at perhaps a hundred universities and a limited number of commercial operations. But most of the technical world still used Fortran and ran on operating systems from hardware vendors, like IBM's System/360. Locally, most programmers at Murray Hill used the GE 635 with GE's batch operating system GECOS (renamed GCOS when GE sold its computer business to Honeywell in 1970).

By 1973, I had started to program in C regularly, but I was still writing Fortran as well. Although Fortran was fine for numerical computation, its control-flow statements were almost non-existent and it was constrained by its origin as a punchcard-based language from the 1950s. By contrast, C control flowed naturally, so to speak.

Accordingly, I wrote a simple compiler that would translate a dialect of Fortran that looked like C into valid Fortran. I called it *Ratfor*, for "rational Fortran." Ratfor converted C control flow, with *if-else*, *for*, *while*, and braces for grouping, into Fortran's IF and GOTO statements and its one looping construct, the DO loop. The preprocessor also provided notational conveniences like free-form input (not the rigidly formatted 80-column card images that Fortran still required), a convenient comment convention, and natural logical and relational operators like < and >= instead of Fortran's klunky .LT. and .GE. forms.

As a short example, here's the Fortran program of Chapter 1 in one of several ways that it might be written in Ratfor:

```
# make v an identity matrix
do i = 1, n
   do j = 1, n
      if (i == j)
         v(i,j) = 1.0
      else
         v(i,j) = 0.0
```

Ratfor was the first example of a language that based its syntax on C. Writing Fortran in Ratfor was, if I do say so myself, infinitely more pleasant than writing standard Fortran. Ratfor didn't change Fortran semantics or data types—it had no features for processing characters, for instance—but for anything where Fortran would be a good choice, Ratfor made it better. Free-form input and C-like control flow made it feel almost like writing in C.

In a tour de force of both theory and practice, Brenda Baker created a program called struct that translated arbitrary Fortran programs into Ratfor. Brenda showed that almost any Fortran program has a well-structured form; there is essentially a unique best way to render it in Ratfor. People who used struct found that the Ratfor version was almost always clearer than the Fortran they had originally written.

Bill Plauger and I decided to write a book to evangelize the Unix tools philosophy for a wider audience: programmers who were writing Fortran on non-Unix systems. *Software Tools*, which was published in 1976, presented Ratfor versions of standard Unix tools: file comparison, word counting, grep, an editor like ed, a roff-like text formatter, and a Ratfor preprocessor itself, all written in Ratfor.

Our timing was about right; the book sold moderately well and a Software Tools User Group sprung up, spearheaded by Debbie Scherrer, Dennis Hall and Joe Sventek at Lawrence Berkeley Lab. They polished and refined the programs, added new tools of their own, made distributions of the code, organized conferences, and kept everything running smoothly for years. Their code was ported to more than 50 operating systems and the user group remained active and influential until it disbanded in the late 1980s.

In 1981 Bill and I published a version of the *Tools* book based on Pascal, which at the time was popular as a teaching language in universities. Pascal had good properties, including sensible control flow and recursion, both of which were missing from Fortran.

Unfortunately, it also had some not so good properties, like awkward input and output and almost unusable character strings, which I described in a paper titled "Why Pascal is Not My Favorite Programming Language." I submitted the paper to a journal, but it was rejected as too controversial, or perhaps not substantial enough. It never was published, but in spite of that, it's cited surprisingly often.

In any case, as C and Unix became more widely available, Pascal's serious limitations made it less and less popular, so *Software Tools in Pascal* was not widely read. In hindsight, a C version would have had far more impact, both in the short term and in the long run.

4.9 Biography: Doug McIlroy

Rob Pike has called Doug McIlroy "the unsung hero of Unix," and I agree. Ken Thompson says that Doug is smarter than everyone else, which also seems accurate, though Doug himself says "I'll leave to others to assess how smart I may be, but I know that many of BTL's practicing mathematicians were much smarter." Suffice it to say that there were many outstanding people at the Labs, the imposter syndrome was not unknown, and one was continuously stretched trying to keep up.

No matter who is right here, Unix might not have existed, and certainly would not have been as successful, without Doug's good taste and sound judgment of both technical matters and people.

Doug received his undergraduate degree in physics from Cornell in 1954 and his PhD in applied mathematics from MIT in 1959. He too worked for a summer at Bell Labs, joined permanently in 1958, and became head of the Computing Techniques Research department in 1965, two years before I first met him. As described earlier, I spent the summer of 1967 as an intern in Doug's department, nominally working on a storage allocator problem that he suggested but in fact doing my own thing. One of his many good qualities as a manager was that he wasn't bothered by that at all.

I've already mentioned Doug's early language work on PL/I and EPL. Once Unix was underway, he wrote a wide variety of fundamental software. His storage allocator `malloc` was used for many years, and his research on allocators affected subsequent work. He also wrote a bunch of commands; the list on his web page at Dartmouth includes `spell`, `diff`, `sort`, `join`, graph, speak, tr, tsort, calendar, echo, and tee.

Some of these are small, like echo, while some are large, like sort and diff, but most are central to Unix computing and many are used to this day. Of course pipes were his idea too, though the final version used Ken's syntax, and their existence was due to Doug's ongoing lobbying for such a mechanism.

His version of spell made effective use of a dictionary and heuristics for identifying parts of speech to find potential spelling mistakes, using only meager resources.

Doug's version of diff implements an efficient algorithm (invented independently by Harold Stone and Tom Szymanski) for comparing two text files, finding a minimal sequence of changes that will convert one into the other. This code is at the heart of source code control systems that manage multiple versions of files. Such systems most often work by storing one version and a set of diffs, generating other versions by running the diff algorithm. This is also used in the patch mechanism that's used for updating programs—rather than sending a new version of a program to someone, send a sequence of ed editing commands, computed by diff, that will convert the old version into the new.

The diff program is another nice example of how good theory can combine with good practical engineering to make a fundamental tool. The output that diff produces is readable by both people and programs; output written for either one or the other would be far less useful. It is also a simple example of a program that writes a program, and its output is a fine little language.

Quite early in Unix days, 1127 bought a novel device, a Votrax voice synthesizer that converted phonemes into sound. Doug created a set of rules for converting arbitrary English text into phonemes, and wrote a program called speak that used the rules to generate Votrax input. English spelling is of course notoriously irregular, so no set of rules can do a perfect job. The output of speak was often imperfect, sometimes funny (my name rhymed: "Br-I-an Kern-I-an"), but almost always accurate enough to be genuinely useful.

The program was just another command, so it could be used by anyone without prearrangement; text sent to speak was played through a loudspeaker in the Unix room. This led to any number of odd services. For instance, every day at 1PM, the Votrax would say

"Lunchtime, lunchtime, lunchtime. Yummy, yummy, yummy."

as a reminder to the inhabitants that it was time to head off to the cafeteria before it closed at 1:15.

There was also a service that monitored incoming telephone lines and when one rang (silently) would announce

"Phone call for Doug"

or whoever; in a shared space it was much less distracting than frequently ringing phones.

Doug's interests were broad and deep. Among other things, he was an expert in map projections, which are a specialized form of mathematics. His `map` program provided several dozen projections, and to this day, he's still producing new ones that show up in Christmas cards sent to friends and on his Dartmouth web page (Figure 4.6).

Figure 4.6: One of Doug McIlroy's many maps

Doug was a superb technical critic, often the first person to try out some new program or idea. He would experiment at the earliest possible time and, since he had excellent taste, his opinions on what was good and what needed to be fixed was invaluable. There was a steady flow of people to his office to get advice and critical comments on ideas, algorithms, programs, documents—pretty much anything. Bjarne Stroustrup used to drop in on me to discuss C++ and explain some new idea, then move a few doors along the corridor to Doug's office to get serious feedback on the language design.

Figure 4.7: Doug McIlroy and Dennis Ritchie, May 2011 (Wikipedia)

Doug was usually the first person to read drafts of papers or manuals, where he would deftly puncture rhetorical balloons, cut flabby prose, weed out unnecessary adverbs, and generally clean up the mess. In Mike Mahoney's oral history of Unix (1989), Al Aho says of Doug,

> "He understood everything that I was working on, with just the most fragmentary descriptions. And he essentially taught me to write, too. I think he's one of the finest technical writers that I know of. He has a flair for language, and a flair for economy of expression that is remarkable."

Doug was the outside reader on my thesis, where he greatly improved the organization and exposition. He also read multiple drafts of all the books that I co-authored with others at the Labs, always making them better. He refined and polished the manual pages for commands, and he pulled together and organized the contents of the manuals for the 8th through 10th editions of Unix. He did all of this with enthusiasm and care, at some cost to his own research.

Doug stepped down from his management role in 1986 and retired from the Labs in 1997, heading off to teach at Dartmouth. Figure 4.7 is a picture taken at Bell Labs in Murray Hill during a 2011 celebration of the Japan Prize awarded to Ken and Dennis.

Chapter 5

Seventh Edition (1976-1979)

> "It was really with v7 that the system fledged and left the research nest. V7 was the first portable edition, the last common ancestor of a radiative explosion to countless varieties of hardware. Thus the history of v7 is part of the common heritage of all Unix systems."
> Doug McIlroy, *A Research Unix Reader*, 1986.

The 6th edition of Unix was a great base for software development, and the tools that came with it made programming fun and productive. Some pieces were in place well before the 6th edition, while others came along afterwards. In this chapter, we'll see several threads of software development in 1127 that culminated in the 7th edition, which was released in January, 1979, nearly four years after the 6th edition.

Logically and chronologically, some of this material should come after the next chapter, which describes the spread of Unix outside of Center 1127, but the story seems more cohesive if I talk about the 7th edition first. And as Doug McIlroy noted in the epigraph above, the 7th edition was the source of much of the heritage shared by all Unix systems.

One of the major themes of Unix and of this chapter is the flowering of influential languages, some aimed at conventional programming, some special-purpose or domain-specific, and some declarative specification languages. I'll probably spend more time on this topic than many readers may care about, but it's been my area of interest for many years. I'll try to describe the important bits early in each section so that you can safely skip to the end.

It's also worth noting that 6th edition Unix was strictly a PDP-11 operating system at the beginning of this period; by 1979, the 7th edition was a portable operating system that ran on at least four different kinds of processors, of which the DEC VAX-11/780 was the most popular. I'll have more to say about portability in the next chapter, but for now it's important to notice that Unix had quietly evolved from being a PDP-11 system to one that was comparatively independent of specific hardware.

5.1 Bourne shell

I/O redirection and pipes in the 6th edition shell made it easy to combine programs to do some task, originally by typing a sequence of commands and then collecting them in a file—a shell script—so they could be run as a single command.

The 6th edition shell had an *if* statement for conditionally executing a command, a *goto* statement for branching to another line of a script file, and a way to label a line in a script (the ":" command, which did nothing) so it could be branched to. Taken together, these could be used to make loops so in principle the 6th edition shell could be used to write complicated scripts. In practice, however, the mechanisms were awkward and fragile.

As I'll describe in the next chapter, John Mashey, a member of the Programmer's Workbench (PWB) group, had added a number of features to his version of the 6th edition shell that made it much better for programming: a general *if-then-else* statement for testing conditions, a *while* statement for looping, and variables for storing information within a shell file.

In 1976, Steve Bourne, who had recently joined 1127, wrote a new shell that incorporated the PWB shell features, along with major enhancements of his own. The goal was to retain the easy interactive nature of the existing shell, but also make it a fully programmable scripting language. Steve's shell provided several control-flow constructs, including *if-then-else*, *while*, *for* and *case*. It also included variables, some of which were defined by the shell itself and others that could be defined by users. Quoting mechanisms were enhanced. Finally it was modified so it could be a filter in a pipeline just like any other program. The result, which was simply called `sh`, quickly replaced the 6th edition shell.

The control-flow syntax of the new shell was unusual, since it was based on Algol 68, a language favored by Steve though not many others in 1127. For example, Algol 68 used reversed words as terminators, like `fi` to terminate `if` and `esac` to terminate `case`. But since `od` was already taken (for the octal dump command), `do` was terminated by `done`.

```
for i in $*        loop over all arguments
do
  if grep something $i
  then
     echo found something in $i
  else
     echo something not found in $i
  fi
done
```

The conditions tested by the `if` and `while` statements were the status returns from programs, that is, numeric values that a program could use to report results like whether it had worked properly. Most programs at the time were cavalier about returning sensible status, since it had rarely mattered before. So Steve set his shell to print an irritating message each time a program didn't produce a sensible status. After a week of this automated nagging, most programs had been upgraded to return meaningful status values.

Steve's shell also significantly enriched I/O redirection. The BUGS section of the 6th edition shell had said "There is no way to redirect the diagnostic output." One especially useful new shell feature was a way to separate the standard error stream (by default file descriptor 2) from the standard output (file descriptor 1), so that the output of a script could be directed to a file while error messages went somewhere else, usually the terminal:

```
prog >file          # stdout to file, stderr to terminal
prog 2>err          # stdout to terminal, stderr to err
prog 1>file 2>err   # stdout to file, stderr to err
prog >file 2>&1     # merge stderr with stdout
```

By this point, the shell had become a real programming language, suitable for writing pretty much anything that could reasonably be formulated as a sequence of commands. It could often do this well enough that there was no need to write a C program.

Over the years, more features have been added, particularly in Bash, the "Bourne Again Shell" that is now the de facto standard shell for most Linux and macOS users. Although personal shell scripts tend to be small and simple, the source code for major tools like compilers is often distributed with configuration scripts of 20,000 or more lines. These scripts run programs that test properties of the environment, like the existence of libraries and sizes of data types, so they can compile a version that has been tuned to the specific system.

5.2 Yacc, Lex, Make

We use language to communicate, and better languages help us to communicate more effectively. This is especially true of the artificial languages that we use to communicate with computers. A good language lowers the barrier between what we want to say ("just do it") and what we have to say to get some job done. A great deal of research in computing is concerned with how to create expressive languages.

Seventh edition Unix offered a diversity of language-based tools, some rather unconventional. I think it's fair to say that the majority of those languages would not exist had it not been for tools, especially Yacc, that made it easy for non-experts to create new languages. This section describes the language-building tools. The overall message is that Unix tools facilitated the creation of new languages and thus led to better ways to communicate with computers. You can safely skip the details, but the message is important.

Computer languages are characterized by two main aspects, syntax and semantics. Syntax describes the grammar: what the language looks like, what's grammatically legal and what's not. The syntax defines the rules for how statements and functions are written, what the arithmetic and logical operators are, how they are combined into expressions, what names are legal, what words are reserved, how literal strings and numbers are expressed, how programs are formatted, and so on.

Semantics is the meaning that is ascribed to legal syntax: what does a legal construction mean or do. For the area computation program in Chapter 2, which is repeated here:

```
void main() {
    float length, width, area;
    scanf("%f %f", &length, &width);
    area = length * width;
    printf("area = %f\n", area);
}
```

the semantics say that when the function `main` is called, it will call the function `scanf` to read two data values from the standard input, compute the area, and call `printf` to print `area =`, the area and a newline character (\n).

A compiler is a program that translates something written in one language into something semantically equivalent in another language. For example, compilers for high-level languages like C and Fortran might translate into assembly language for a particular kind of computer; some compilers translate from other languages, such as Ratfor into Fortran.

The first part of the compilation process is to *parse* the program, that is, to determine its syntactic structure by recognizing names, constants, function definitions, control flow, expressions, and the like, so that subsequent processing can attach suitable semantics.

Today, writing a parser for a programming language is well-understood technology, but in the early 1970s it was an active research area, focused on creating programs that would convert the grammar rules of a language into an efficient parser for programs written in that language. Such parser-generator programs were also known as "compiler-compilers," since they made it possible to generate the parser for a compiler mechanically. Typically they created a parser and also provided a way to execute code when particular grammar constructs were encountered during parsing.

Yacc

In 1973, Steve Johnson (Figure 5.1), with language-theory help from Al Aho, created a compiler-compiler that he called YACC (henceforth Yacc). A comment from Jeff Ullman inspired the name, which stands for "yet another compiler-compiler," suggesting that it was not the first such program.

Figure 5.1:   Steve Johnson, ~1984 (Courtesy of Gerard Holzmann)

A Yacc program consists of grammar rules for the syntax of a language, and semantic actions attached to the rules so that when a particular grammatical construction is detected during parsing, the corresponding semantic action can be performed. For example, in pseudo-Yacc, part of the syntax of arithmetic expressions might be:

```
expression := expression + expression
expression := expression * expression
```

and the corresponding semantic actions might be to generate code that would add or multiply the results of the two expressions together and make that the result. Yacc converts this specification into a C program that parses input and performs the semantic actions as the input is being parsed.

Normally a compiler writer would have to write more complicated rules to handle the fact that multiplication has higher precedence than addition (that is, multiplications are done before additions), but in Yacc, operator precedence and associativity can be specified by separate declarations rather than additional grammar rules, which is a huge simplification for non-expert users.

Steve himself used Yacc to create a new "portable C compiler" (PCC) that had a common front end for parsing the language and separate back ends for generating code for different computer architectures. Steve and Dennis also used PCC in their implementation of Unix for the Interdata 8/32, as described in Section 6.5.

PCC had other uses as well. As Steve recalls,

"An unexpected spin-off from PCC was a program called Lint. It would read your program and comment on things that were not portable, or just plain wrong, like calling a function with the wrong number of arguments, inconsistent sizes between definition and use, and so on. Since the C compiler only looked at one file at a time, Lint quickly became a useful tool when writing multi-file programs. It was also useful in enforcing standards when we made V7 portable, things like looking for system calls whose error return was −1 (Version 6) instead of null (V7). Many of the checks, even the portability checks, were eventually moved into the C language itself; Lint was a useful testbench for new features."

The name *Lint* comes from the image of picking lint off clothing. Although its functionality is now often subsumed into C compilers, the idea is common in analogous tools for a number of other languages.

Yacc was instrumental in several of the languages that were developed in 1127 over the years, some of which are described in the next few sections. Lorinda Cherry and I used it for the mathematical typesetting language Eqn. I also used Yacc for the document preparation preprocessors Pic and Grap (the latter with Jon Bentley, Figure 5.2), for the AMPL modeling language, for at least one version of Ratfor, and for other one-off languages over the years. Yacc was also used for the first Fortran 77 compiler `f77`, Bjarne Stroustrup's C++ preprocessor `cfront`, the Awk scripting language (to be described shortly), and a variety of others.

Figure 5.2: Jon Bentley, ~1984 (Courtesy of Gerard Holzmann)

Yacc's combination of advanced parsing technology, high efficiency and convenient user interface helped it to become the sole survivor among the early parser generators. Today it lives on under its own name, in independent implementations like Bison that are derived from it, and in reimplementations in half a dozen other languages.

Lex

In 1975, Michael Lesk (Figure 5.3) created a lexical analyzer generator called Lex, which is a direct parallel to Yacc. A Lex program consists of a sequence of patterns (regular expressions) that define the "lexical tokens" that are to be identified; for a programming language, these would be components like reserved words, variable names, operators, punctuation, and so on. As with Yacc, a semantic action written in C can be attached to each

Figure 5.3: Michael Lesk, ~1984 (Courtesy of Gerard Holzmann)

specified token. From these, Lex generates a C program that will read a stream of characters, identify the tokens it finds, and perform the associated semantic actions.

Mike wrote the first version of Lex but it was quickly revised in the summer of 1976 by an intern who had just graduated from Princeton. Mike recalls:

> "Lex was rewritten almost immediately by Eric Schmidt as a summer student. I had written it with a non-deterministic analyzer that couldn't handle rules with more than 16 states. Al Aho was frustrated and got me a summer student to fix it. He just happened to be unusual."

Eric went on to a PhD at Berkeley, and was the CEO of Google from 2001 to 2011.

Yacc and Lex work well together. Each time Yacc needs the next token while parsing, it calls on Lex, which reads enough input to identify a complete token and passes that back to Yacc. The Yacc/Lex combination mechanizes the front-end components of a compiler while taking care of complicated grammatical and lexical constructs. For example, some programming languages have operators that are two or three characters long, like the ++ operator in C. When the lexical analyzer sees a +, it has to look ahead to know whether the operator is ++ or an ordinary + followed by something else. It's not too hard to write this kind of code by hand, but it's a lot easier to have it written for you. In Lex, one would only have to say

```
"++" { return PLUSPLUS; }
"+"  { return PLUS; }
```

to distinguish the two cases. (`PLUS` and `PLUSPLUS` are names for numeric codes that are easy for the C code to deal with.)

Figure 5.4 shows how Yacc and Lex are used in the creation of a C program that is a compiler for some language. Yacc generates a C file for the parser and Lex generates a C file for the lexical analyzer. These two C files are combined with other C files that contain semantics, and compiled by a C compiler to make an executable program. This figure was created with Pic, which has exactly this structure.

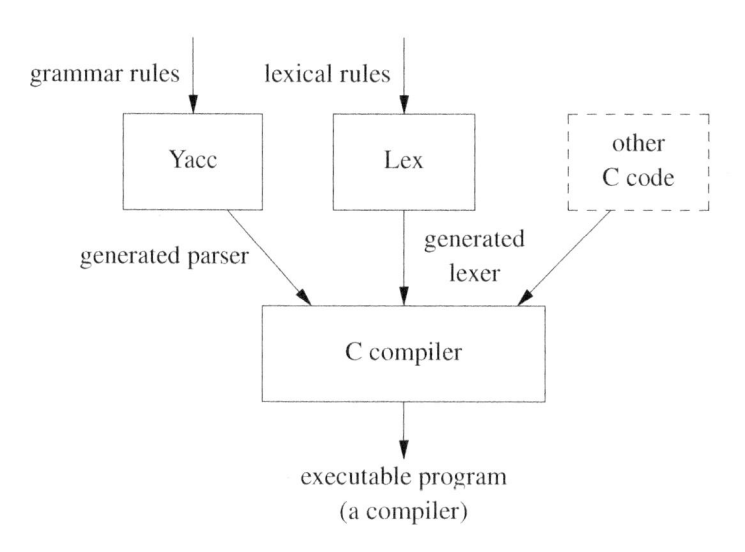

Figure 5.4: Using Yacc and Lex to create a compiler

In spite of how easy and powerful Lex is, over the long haul, it has not been as extensively used as Yacc has. Perhaps this is because writing a parser for a complex language can be daunting for a comparatively inexperienced programmer, while writing a lexical analyzer is not. But writing a lexical analyzer by hand, however easy and straightforward it might seem, is not necessarily a good idea.

My experience with the Awk scripting language (discussed later in this chapter) may be instructive. The first implementation of Awk used Yacc for the grammar and Lex to tokenize the input program. When we tried to port Awk to non-Unix environments, however, Lex was not available or it worked differently. After a few years, I reluctantly rewrote the lexical part of Awk in C, so it would be portable to all environments. But for years afterwards, that hand-crafted lexical code was a fruitful source of bugs and subtle problems that were not present in the Lex-generated version.

This is a good example of a general rule: if a program writes your code for you, the code will be more correct and reliable than if you write it yourself by hand. If the generator is improved, for example to produce better code, everyone benefits; by contrast, improvements to one hand-written program do not improve others. Tools like Yacc and Lex are excellent examples of this rule, and Unix provides many others as well. It's always worth trying to write programs that write programs. As Doug McIlroy says, "Anything you have to do repeatedly may be ripe for automation."

Make

Most large programs consist of multiple source files, which have to be compiled and linked together to create an executable program. This can often be done with a single command such as `cc *.c` to compile all the source files for a C program, but in the 1970s computers were so slow that recompiling a multi-file program after making a change in a single file could take significant time, minutes instead of seconds. It was more efficient to recompile only the changed file and link the result with the other previously compiled files.

Remembering which files had been compiled recently enough and which needed to be recompiled was a nuisance, however, and it was easy to make mistakes. Steve Johnson complained about this to Stu Feldman (Figure 5.5) one day, after spending hours of fruitless debugging, only to realize that he had simply failed to recompile one of the files he had changed.

Coincidentally, Stu had done exactly the same thing and had also struggled to debug a program that was really correct, just not recompiled. He came up with a elegant idea, a specification language that describes how the pieces of a program depend on each other. A program that he called Make analyzed the specification and used the times that files had been changed to do the minimum amount of recompilation necessary to bring everything up to date. The first implementation was in 1976:

> "I wrote Make over the weekend, and then rewrote it the next weekend with macros (the list of built-in code was getting too long). I didn't fix the tab-in-column-1 because I quickly had a devoted user base of more than a dozen people and didn't want to upset them."

Figure 5.5: Stu Feldman, ~1984 (Courtesy of Gerard Holzmann)

Make was an instant success, since it obviated a whole class of silly errors while making compilation as efficient as possible. It was also a boon for programs that involved more complicated processing than just C compilation, for example programs that used Yacc and Lex, each of which had to be run first to create C files that were then compiled, as in Figure 5.4 above. One makefile could capture all the processing steps necessary to compile a new version of a program, and could also describe how to do related tasks like running Lint, making a backup and printing documentation. A makefile had some of the same properties as a shell script, but the language was declarative: a specification of dependencies and how to update components but without explicit tests of file creation times.

The "tab-in-column-1" problem that Stu refers to is an unconventional and somewhat awkward restriction on the format of makefiles. It's arguably a design flaw. It's also a nice example of a general problem that any successful program faces: if the program is good, it attracts users, and then it becomes hard to change the program in any incompatible way. Unix and most other systems are replete with examples of initial blemishes that are now too entrenched to fix.

Make is also a good example of a theme in this section: rather than writing code or doing sequences of operations by hand, create a notation or specification that declares what has to be done, and write a program to interpret the specification. This approach replaces code with data, and that's almost always a win.

Yacc, Lex and Make are very much with us today, because they address important problems that programmers still face, and they solve those problems well enough that the designs and sometimes even the original implementations are still in use.

As a digression, I first met Stu in about 1967 when I was a grad student at Princeton and he was an undergrad; he was working part-time on Multics for Bell Labs during the school year. He joined 1127 after getting his PhD in astrophysics at MIT. He went to Bellcore in 1984, then to IBM, then to Google, where in a bit of good fortune for me, he was my manager several levels up when I visited there for summers.

5.3 Document preparation

Unix had good tools for document production from very early on, and this helped to make its documentation good. This section tells an extended story about the history of document preparation tools on early Unix systems. Like so much of Unix, it's a story of how the interactions among programs, programmers and users formed a virtuous cycle of innovations and improvements.

When I was an intern at MIT in 1966, I encountered Jerry Saltzer's Runoff program. (The name comes from expressions like "I'll run off a copy for you.") Runoff was a simple text formatter: its input consisted of ordinary text interspersed with lines beginning with a period that specified formatting. For example, a document might say

```
.ll 60
.ce
Document preparation
.sp 2
.ti 5
Unix had good tools for document production ...
.sp
.ti 5
When I was an intern at MIT in 1966 ...
```

This "markup" told Runoff how to format the text: set the line length to 60 characters, center the next line, space down two lines, temporarily indent 5 spaces, print the paragraph in lines of at most 60 characters, then space down one line and temporarily indent again for the next paragraph.

Runoff had a dozen or two commands like this that made it easy to format simple documents—manual pages, program descriptions, letters to friends—any text formatting that one might do today with a tool like Markdown.

Early formatters

Runoff was a revelation to me, a way to use computers that had nothing to do with mathematical computations or compiling. It became easy to refine one's writing over and over again at little cost. It may be hard for readers today to appreciate just how labor-intensive it was to prepare documents before the creation of word processing programs, when there were only mechanical typewriters—better than clay tablets or quill pens, to be sure, but any change of more than a few words in a document would require a complete retype. Thus most documents went through only one or two revisions, with handwritten changes on a manuscript that had to be laboriously retyped to make a clean copy.

When I started to write my thesis in the fall of 1968, I really wanted Runoff, since the alternative would have been to type the thesis myself on a manual typewriter (and retype it for each set of changes), or pay someone to do it for me. I'm a fast but inaccurate typist, so the former was not practical, and since I was both cheap and poor, the latter wasn't either.

Thus I wrote a simple version of Runoff that I called "Roff," for "an abbreviated form of Runoff." The problem was that there was no interactive computer system like CTSS at Princeton; there weren't any computer terminals either. All that was available was punch cards, which only supported upper case letters. I wrote Roff in Fortran (far from ideal, since Fortran was meant for scientific computation, not pushing characters around, but there were no other options) and I added a feature to convert everything to lower case while automatically capitalizing the first letter of each sentence. The resulting text, now upper and lower case, was printed on an IBM 1403 printer that could print both cases. Talk about bleeding edge! My thesis was three boxes of cards. Each box held 2,000 cards, was about 14 inches (35 cm) long and weighed 10 pounds (4.5 kg). The first 1,000 cards were the program and the other 5,000 were the thesis itself in Roff.

Readers who have never worked with cards may find this confusing. Each card contained at most 80 characters, either one line of Fortran code or one line of thesis text. If some part of the text needed to be changed, the replacement text was punched onto a few new cards that replaced the old cards, which were discarded. Fixing a spelling mistake would generally require replacing only one card, though if the new text were much longer, more cards might be needed.

$$\sum_{j=1}^{m} c[p(j),k] = \sum_{j=1}^{r} c[q(j),k]$$

This follows from the fact that for any i, the cost $c[q(i),k]$ is allocated among that subset of the $p(j)$'s which are copies of $q(i)$. That is, $\Sigma c[p(j),k] = c[q(i),k]$ for any such subset. Summation of this equality over all $q(i)$ proves the claim.

By construction, the cost for edges leaving the i-th copy of node k in the derived tree is

$$c[k(i),k'(i)] = c(k,k') \frac{c[p(i),k]}{\sum_{j=1}^{m} c[p(j),k]}$$

But

$$\sum_{j=1}^{m} c[p(j),k] = \sum_{j=1}^{r} c[q(j),k] \leq c(k,k')$$

Therefore

$$c[k(i),k'(i)] \geq c[p(i),k]$$

and hence monotonicity of subroutine graph costs is preserved in the tree. Equality of values of edges leaving a copy of a particular node is obviously preserved since the same multiplying factor is used for all the edges leaving the given node.

Figure 5.6: A page of my thesis, formatted with Roff

I had to manually insert a few special characters like summation signs (Σ) into the printed pages but this kludgy mechanism worked surprisingly well, certainly enough for me to print my thesis, which I believe was the first computer-printed thesis at Princeton. (Figure 5.6 shows a random page.) For some years afterwards, there was a student agency that would "roff" documents for students for a modest fee. Roff was thus the first program I ever wrote that was used by other people in any significant way.

When I got to Bell Labs, I found that there were a couple of other roff-like programs also underway, including one by Doug McIlroy that was based on Saltzer's original. And Joe Ossanna shortly thereafter wrote a much more powerful version that he called Nroff, for "new Roff," which made it possible for the patent department to format patent applications. As I described earlier, Nroff was the critical tool that enabled the purchase of the first PDP-11 computers for Unix.

This little pocket of document preparation enthusiasts, and a community of active users of such programs, fit my interests perfectly, and I spent a significant part of the next ten years happily working on tools for text formatting.

Troff and typesetters

Roff and Nroff only handled fixed-width ("monospace") character sets, not much more than the standard alphabetic characters found on the Model 37 Teletype, so the output quality wasn't very high. In 1973, however, Joe Ossanna arranged to buy a phototypesetter, a Graphic Systems CAT, which was popular in the newspaper industry. Joe's intent was to produce better-looking internal technical documents and also to help the patent department to prepare better patent applications.

The CAT could print conventional proportionally spaced fonts in roman, italic and bold, along with a set of Greek letters and special characters for mathematics. It printed on long rolls of photographic paper that had to be developed in a couple of baths of noxious and messy chemicals. This technology predates laser printers, which did not become widely available for at least another 10 years. Furthermore, the output was black and white; inexpensive color printing did not arrive until several decades later.

Each font was a piece of 35mm film with character images, mounted on a rapidly spinning wheel. The wheel held four fonts of 102 characters each, so the total repertoire for a single job was 408 characters. The typesetter flashed an intense light through the film strip image onto photographic paper when the paper and the desired character were in the right positions. It was capable of 16 distinct sizes.

The typesetter was slow—changing sizes required it to rotate a mechanical lens turret—and the photographic chemicals were most unpleasant, but the output quality was high enough that we could produce professional-looking documents. Indeed, there were occasions when a paper submitted to a journal by a Bell Labs author was questioned: it looked so polished that it must have already been published.

To drive the typesetter, Joe created a significant extension of Nroff that he called Troff. "T" is for typesetter; it's pronounced tee-roff. The Troff language was arcane and tricky, but with sufficient skill and patience, it could be made to do any formatting task, though few people ever mastered it. In effect, Troff was an assembly language for a truly weird computer, so most people used it through packages of macros that encapsulated common formatting operations like titles, headings, paragraphs, numbered lists, and so on. The macros provided a higher-level language above the low-level Troff commands. Mike Lesk, who created the widely used `ms` package, was the master of creating macro packages; no one else in my orbit came close to his skill in using the programming capabilities of Troff.

Once we had a typesetter that could produce output in a variety of fonts with proportional spacing and enough special characters, it became possible to think about typesetting books as well as internal technical documents. The first book produced in that way was *The Elements of Programming Style*, which Bill Plauger and I wrote in 1974. It was typographically rough in many places because we did not have a monospace font for displaying programs, but it was otherwise satisfactory.

One of the main motivations that Bill and I had for doing our own typesetting was to avoid the errors that the conventional publishing process frequently introduced into printed computer programs. Because we had total control over our content, from input to final pages ready to be printed, we could test the programs directly from the text, which would never be touched by copy-editor or compositor hands. The result was an essentially error-free programming book, which was most unusual at the time. I've used that same process ever since; the books listed at the front of this one have all been produced with Troff or its modern incarnation, Groff. Fortunately, one no longer needs typesetters and their expensive and unpleasant media. Today it's sufficient to get everything right in a PDF file and send that to a publisher or printer.

Eqn and other preprocessors

Bell Labs authors wanted to create documents that contained more than just text, most obviously mathematics, but also tables, figures, bibliographic citations, and so on. Troff itself was capable in principle of handling such things, but not in any remotely convenient way. Thus we began to create special-purpose languages that made it easier to handle specific types of technical material. In effect, we were repeating for document preparation the kind of evolution that had already happened for conventional programming languages.

The first of these special-purpose languages was Eqn, a language and program for typesetting mathematical expressions that Lorinda Cherry (Figure 5.7) and I wrote in 1974. As might be expected of a scientific research lab, Bell Labs produced a great number of technical documents, mostly for internal consumption, and many of those were full of mathematics. The Labs had a cadre of talented typists who could read hand-written mathematics and type it into recognizable form using manual typewriters, but this process was time-consuming and edits were painful.

Figure 5.7: Lorinda Cherry, ~1984 (Courtesy of Gerard Holzmann)

Lorinda had been exploring the idea of a tool for printing mathematics, and I wanted a language that would match the way that mathematics is spoken aloud by mathematicians. I think that this language idea was in my subconscious, because while I was a graduate student, I had volunteered for several years to read technical books aloud so they could be recorded onto audio tapes at Recording for the Blind, and so I had spent many hours speaking mathematics.

Eqn did a decent job on simple mathematical expressions. For example, the summation

$$\sum_{i=0}^{\infty} \frac{1}{2^i} = 2$$

is written as

```
sum from i=0 to inf 1 over 2 sup i = 2
```

Eqn proved to be easy to teach to the mathematical typists and then to others, and experiments verified that it was much faster than manual typewriters. The language was simple enough that PhD physicists could also learn it, and after a while, people started doing their own typing rather than relying on expert typists. Eqn was one of the inspirations for math mode in Don Knuth's TeX (1978), which is now the standard for mathematical typography.

Eqn was implemented as a preprocessor for Troff; the normal usage was to pipe the output of Eqn into Troff like this:

```
eqn file | troff >typeset.output
```

Eqn recognized mathematical constructs and translated those into Troff commands, while passing everything else through untouched. The preprocessor approach provided a clean separation into two languages and two programs with different concerns. Lorinda and I had been forced into a good idea by the physical limitations of the PDP-11. There simply wasn't enough memory to include mathematical processing in Troff, which was already about as big as a program could be, and in any case, Joe Ossanna would not have encouraged us to modify Troff even if we had wanted to.

The Eqn language is based on a box model: an expression is built up as a sequence of boxes positioned and sized relative to each other. For example, a fraction is a numerator box centered over a denominator box, with a long-enough line between them. A subscripted expression like x_i is a pair of boxes where the contents of the second box are in a smaller size and positioned somewhat down from the first box.

We used Steve Johnson's newly invented Yacc compiler-compiler to define the grammar and hang semantics on it. Eqn was the first Yacc-based language that wasn't a traditional compiler for a traditional language. Speaking for myself, Eqn would not have happened without Yacc, since I was not up to writing a parser by hand for a new language. The grammar was too complicated to write an *ad hoc* parser for, and it changed frequently while Lorinda and I were experimenting with syntax. Our experience with Yacc is a compelling example of how having good tools makes it possible to do things that would otherwise be too hard or not even conceivable.

Preprocessors for different kinds of typographically difficult material turned out to be a good idea. Soon after Eqn, Mike Lesk created Tbl, which provided a very different language for specifying complex tables, and Refer for managing bibliographic citations, which were important for technical papers.

Many of the programs described in this chapter are preprocessors, that is, programs that convert some language into a suitable form for a subsequent process. Cfront, the original version of C++, would be more accurately described as an object-oriented preprocessor for C that evolved into C++. Sometimes, as with C++, the preprocessor eventually went away as functionality was absorbed into the downstream process. Most other times, however, the programs stayed separate, as with the document preparation tools like Eqn and Tbl. Another example is bc, a preprocessor for dc, Bob Morris's original unlimited-precision calculator. Lorinda Cherry wrote bc to provide conventional infix arithmetic notation for dc, whose postfix notation was hard for novices to use casually.

Preprocessors have many advantages. First, if one implements a language, it is not limited by existing syntax but can use a completely different style, as with the various Troff preprocessors. Second, when memories were small, it was simply not possible to include more functionality in already-large programs; this was especially the case with Troff. Finally, because the output of a preprocessor is available, it can be manipulated before being passed on, to handle other kinds of processing. In the document preparation suite, I've often used Sed scripts and the like to fix character sets and spacing. Chris Van Wyk and I wrote programs to do vertical justification of pages, by modifying the output of Troff before it went to a device driver. These would not have been possible if the functionality were embedded in a single program, but when the process is a pipeline, it's easy to add new stages at the front or back or middle.

Device-independent Troff

Joe Ossanna died in 1977 at the age of 48. Part of his legacy was the source code for Troff, which at the time was nearly 10,000 lines of inscrutable C that Joe had hand-transliterated from its original assembly language form — no comments, dozens of global variables with 2-letter names, and (see the discussion of memory above) a variety of subtle tricks for cramming as much information as possible into not enough memory. In Joe's defense, this was absolutely necessary to pack all of Troff's functionality into 65K bytes, the maximum memory that was available to user programs on the PDP-11/45 that we were using at the time.

I did nothing with the code for at least a year, but finally got up enough courage to start fiddling with it. Slowly and cautiously, I began an upgrade. The biggest single problem, aside from the fact that there were no comments or documentation beyond the user manual, was that it was strongly dependent on the original Graphic Systems CAT typesetter, which by this time was obsolete.

Eventually I managed to find all the places where the code relied on peculiarities of the CAT and replaced them with generic code driven by tables of typesetter characteristics like character sets, sizes, fonts, and resolutions. I invented a typesetter description language so that Troff could produce output that was tuned to the capabilities of a specific typesetter. Straightforward drivers converted that output into whatever input was needed for specific devices. This resulted in the so-called device-independent version called Ditroff. It also enabled other document preparation preprocessors, especially Pic, which was able to take advantage of the higher resolution of new typesetting devices to draw lines and figures.

One of those devices was a new typesetter from Mergenthaler called the Linotron 202. On paper, this device seemed like just what we needed to replace the CAT. It was fast, it had high resolution, it drew characters by painting them on a screen, its processor was a standard minicomputer (a Computer Automation Naked Mini), and it was controlled by a simple program similar to the ones that I had written for other typesetters. The main drawback was the cost, $50,000 in 1979, but we had done enough good work with the previous typesetter that management approved the purchase with hardly any discussion.

Once the 202 arrived, we discovered that its hardware was impossibly flaky; the only thing worse was its software. This started several months of almost daily visits from Mergenthaler's repair service, and a remarkable feat of reverse engineering of the hardware by Ken Thompson and Joe Condon (Figure 5.8).

Joe was a physicist by training, but as his interests shifted, he became an exceptional designer of electronic circuits. He wrote many of the circuit design tools that the Center used for its hardware experiments, and with Ken was the designer of the Belle chess machine. His hardware expertise was crucial to figuring out the 202.

Ken began by writing a disassembler for the binary programs that ran on the machine. (For the record, he did this in a couple of hours one evening, while I went home for dinner and then came back for an evening of work.)

Disassembling Mergenthaler's programs gave Ken and Joe a toehold into how the typesetter itself worked, and after several weeks of intense reverse engineering they figured out Mergenthaler's proprietary encoding of characters and wrote code so we could create our own characters, like the then-current Bell System symbol at the top of Figure 5.9, a chess font for printing

Figure 5.8: Joe Condon, ~1984 (Courtesy of Gerard Holzmann)

games and board diagrams, and a Peter face that had a variety of uses (for example, Figure 5.10).

Ken wrote a B interpreter for the Mergenthaler controller, and we wrote B programs to drive it. This story is told in detail in a technical memo that Bell Labs management suppressed at the time, probably to avoid revealing any of Mergenthaler's intellectual property, but which was finally published in 2013 (see *www.cs.princeton.edu/~bwk/202*). Figure 5.9 shows part of the first page; the incomplete internal memorandum numbers like 80-1271-x indicate that an official number was never assigned.

When it eventually began to work, the high resolution of the 202 made it possible to achieve interesting graphical effects, including half-tone images and line drawings like the diagram of a digital typesetter in Figure 5.9. For the latter, I created a language called Pic in which figures like organization charts or network packet diagrams could be described textually. Naturally it used Yacc for the grammar and Lex for the lexical part. Figure 5.11 shows a simple example of Pic input and output.

Book publication

One reason why the document preparation tools worked well is that they were used for everything—manuals, technical papers, books. If a program had bugs or didn't work well, the authors of the code were in the same

Bell Laboratories

Subject: **Experience with the Mergenthaler Linotron 202 Phototypesetter, or, How We Spent Our Summer Vacation**
Case- 39199 -- File- 39199-11

date: **January 6, 1980**

from: **Joe Condon
Brian Kernighan
Ken Thompson**

TM: **80-1270-1,
80-1271-x,
80-1273-x**

MEMORANDUM FOR FILE

1. Introduction

Bell Laboratories has used phototypesetters for some years now, primarily the Graphic Systems model CAT, and most readers will be familiar with *troff* and related software that uses this particular typesetter.

The CAT is a relatively slow and antiquated device in spite of its merits (low cost, and until recently, high reliability). Most newer typesetters use digital techniques, rather than the basically analog approach of film stencil and optical plumbing used in the CAT. These typesetters store their characters digitally, using some representation of the character outline, and print on photographic paper by painting some area with a CRT. Figure 1 is a block diagram of a typical digital typesetter.

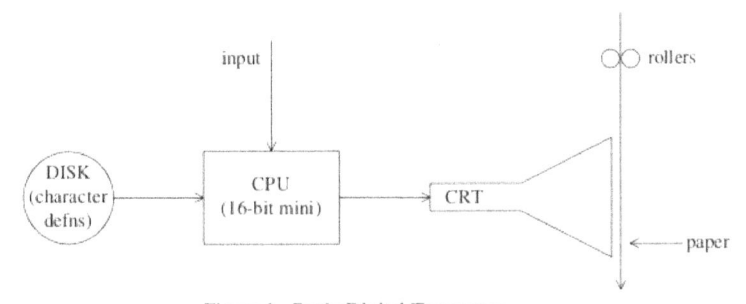

Figure 1: Basic Digital Typesetter

Figure 5.9: Unpublished Bell Labs memo on doing battle with the Linotron 202

hallway, and there was strong pressure to fix things and to add features when necessary. This applied more broadly than just document preparation software, of course—we were all users of our own software, and that gave us a real incentive to improve it.

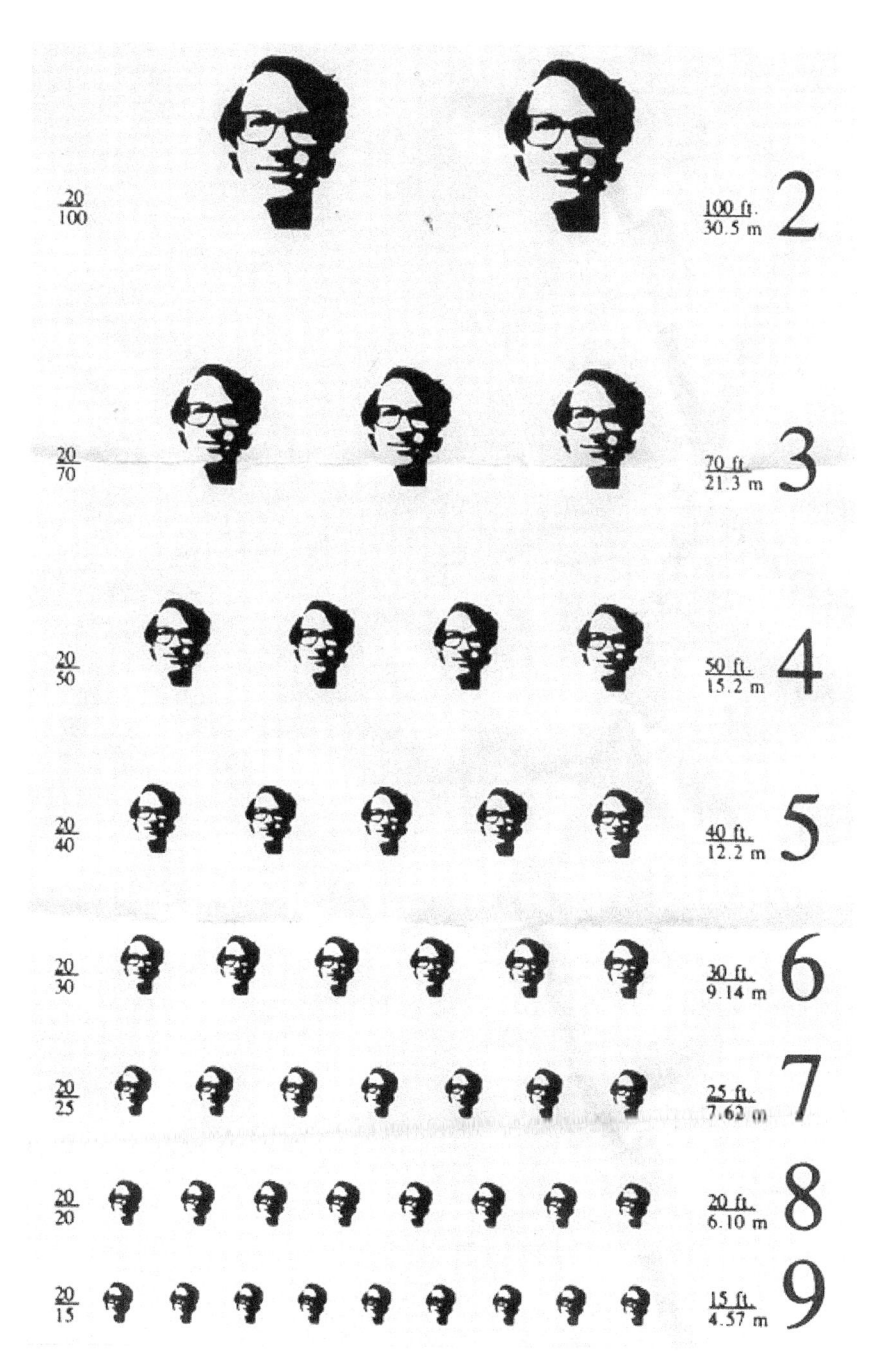

Figure 5.10: Peter face eye-chart (Courtesy of Gerard Holzmann)

```
arrow "stdin" above
box "command"
arrow "stdout" above
arrow down from last box.s "   stderr" ljust
```

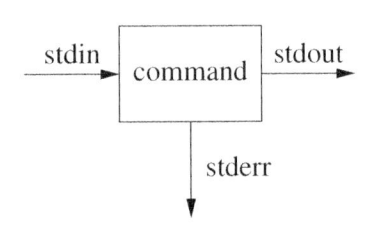

```
           stdin  ┌─────────┐ stdout
          ──────▶ │ command │ ──────▶
                  └────┬────┘
                       │ stderr
                       ▼
```

Figure 5.11: The Pic drawing language (input and output)

Members of the Computing Science Research Center wrote an unusually large number of influential books during the 1970s and 1980s, well beyond what one might expect from an industrial research lab. As a result, after a while, Bell Labs became known as a source of authoritative books about computing and computer science.

Al Aho wrote several widely used textbooks, including the famous 1977 "Dragon book" (*Principles of Compiler Design*, Figure 5.12) with Jeff Ullman, and *Design and Analysis of Computer Algorithms* with Jeff and John Hopcroft. Bjarne Stroustrup (Figure 5.13) created C++ in the early 1980s and wrote several C++ books a few years later. Jon Bentley's *Programming Pearls* books grew out of his columns for *Communications of the ACM*. Mike Garey and David Johnson of the math center used Troff and Eqn for their core book *Computers and Intractability: A Guide to the Theory of NP-Completeness*. We also published book-form manuals for Unix, Plan 9, and so on. These publications became standard texts and references for a generation of programmers and computer science students.

How did this relatively small group of researchers from industry manage to produce so many influential books?

I can see several reasons. First and foremost, people took writing seriously, they took pains with their own writing, and they were great critical readers of what other people wrote. Doug McIlroy was first among this group; no one else matched Doug's ability to spot errors (from tiny to crucial) no matter what the topic, nor had his eye for murky prose. I don't think that I ever wrote anything at Bell Labs that I did not ask Doug to comment on, and he always did. It was humbling when he shredded my words but it made me into a much better writer, and that happened to others as well.

Figure 5.12: Aho and Ullman Dragon book, first edition, 1977

Of course Doug wasn't the only critical reader. Everyone gave generously of their time; it was simply part of the culture that you provided detailed comments on what your colleagues wrote. This was unusual, and was one of the things that made the Labs a great place to be.

Second, local management was supportive of book writing. Publication was important for maintaining Bell Labs' reputation in the scientific and academic communities, and that included books. With managerial support, it was possible to devote full time to a book for six months, and that concentrated effort was enough to basically finish a job that might take several years if it were done as a part-time or evening activity. And it gets better: although Bell Labs retained the copyrights on books, the authors got to keep the royalties. I doubt that any of us ever wrote a book explicitly to make money—no one at the Labs would be dumb enough to think that writing a technical book was lucrative—but if a book had some success, the authors kept the money.

That enlightened management and company policy encouraged people to write, and in the long run it paid off for the company as well as for the authors. Publications by Bell Labs authors were important for recruiting.

Figure 5.13: Bjarne Stroustrup, ~1984 (Courtesy of Bjarne Stroustrup)

Bell Labs wasn't some mysterious secret operation; students knew it as the place where their software was created and their textbooks were written. Potential new hires could see that good work was being done and it was published; they didn't have to worry about disappearing into an "industrial" research lab. That put Bell Labs recruiting on an equal footing with universities, with the added advantage that people could do research full time at the Labs, without the distractions of teaching, administration and raising money. This combination of great software and influential books was a big part of what made the Labs so successful at the time.

A third factor is more technical: the symbiosis among C and Unix as a programming environment, document preparation as a research area, and writing about technical computer topics as a major activity. This began with text formatting programs like Doug McIlroy's Roff, Joe Ossanna's Nroff and Troff, and then the preprocessors like Eqn, Tbl, and so on. Those tools made it easier to produce documents that included ever more typographically challenging content, like mathematics, tables, figures, diagrams, and graphs. That in turn led to better writing, because all of these document preparation programs shared a vital characteristic: they made it easy to make multiple revisions of documents and always have a clean copy to work from, as opposed to the slow and painful alternative of giving material to a typist and waiting days for it to come back.

It may sound trivial, but I'm sure that the ability to make revisions so easily led to better writing, because it pretty much eliminated the overhead of making an up to date copy, and it completely eliminated middlemen like typists, editors and printers. Accuracy mattered for technical papers and for the Unix Programmer's Manual, but control of the whole process especially mattered for books. For programming books, it was vital that the programs were typeset directly from the source code, so we could be sure that what was printed was correct, that it hadn't been inadvertently changed by human intervention.

These tools were all written in C, of course, since it was expressive and efficient. It's hard to remember today, perhaps, that efficiency in both time and space were crucial when machine capacities were expressed in kilobytes, not gigabytes. Every byte counted, and so at some level did every instruction, so a language that economized on both was not just nice, but a practical necessity.

All of this has come full circle—this book was produced by the descendants of these document preparation programs, though with the excellent fresh implementations and enhancements in Groff, Geqn, and so on, written by James Clark.

5.4 Sed and Awk

One of the major simplifications of the Unix file system was its uniform treatment of files as sequences of uninterpreted bytes. There were no records, no required or prohibited characters, and no internal structure imposed by the file system—just bytes.

There was a similar simplification in the way that most Unix programs handled textual data. Text files were just sequences of bytes that happened to be characters in ASCII, the American Standard Code for Information Interchange. This uniform view of plain text was a natural fit for pipelines, and the Unix toolkit was full of programs that read text input, did something with or to it, and wrote text output. I've already mentioned examples like word counting, comparison, sorting, transliteration, finding duplicates and of course the quintessential example, grep for searching.

Sed

The success of grep inspired Lee McMahon to write an analog called `gres` that did simple substitutions on text as it flowed through; "`s`" was the substitute command in `ed`. Lee soon replaced it by a generalization, a stream editor called Sed that applied a sequence of editing commands to text on its way from input to output; grep and gres were both special cases of Sed. The commands that Sed uses are the same as the editing commands in the standard ed text editor. Sed is frequently used today in shell scripts, for transforming a stream of data in some way: replacing characters, or adding or removing unwanted spaces, or discarding something unwanted.

Lee had an unusual background—a PhD in psychology from Harvard, and time in a Jesuit seminary preparing for the priesthood before switching to a more conventional path as a computer scientist. He was one of the first in the Unix group to think about processing text at scale, at a time when primary memories were too small to store large amounts of text. I might add that "large" is relative. Lee's particular interest in the early 1970s was the Federalist Papers, which altogether are only a little over a megabyte.

Awk

I was interested in tools that could process both numbers and text equally well. Neither grep or Sed could handle numeric data or do arithmetic, and grep couldn't deal with multiple text lines; such computations still required a C program. So I was looking for some kind of generalization. At the same time, Al Aho (Figure 5.14) had been experimenting with a richer class of regular expressions than grep supported, and had written egrep ("extended grep"). Finally, Peter Weinberger, who not long afterwards transferred into 1127 and moved into the office between Al and me, was interested in databases.

In the fall of 1977, the three of us talked about how to combine these ideas, taking some inspiration from RPG, IBM's powerful but inscrutable report program generator, along with a neat idea from Marc Rochkind that's described in the next chapter. Ultimately we designed a language that we called AWK (Awk from now on). As we said in the original description, naming a language after its authors shows a certain paucity of imagination. I don't recall now whether we thought about cognates related to "awkward," or perhaps we found the name apt in a sardonic way, but in any case it stuck. Peter wrote the first version very quickly, in just a few days, using Yacc, Lex and Al's egrep regular expression code.

An Awk program is a sequence of patterns and actions. Each line of input is tested against each pattern, and if the pattern matches, the corresponding action is performed. Patterns can be regular expressions or numeric or string relations. Actions are written in a dialect of C. An omitted pattern matches all lines; an omitted action prints the matching line.

Figure 5.14: Al Aho, ~1984 (Courtesy of Gerard Holzmann)

This example prints all input lines that are longer than 80 characters; it's a pattern with no action.

```
awk 'length > 80'
```

Awk supports variables that hold numbers or strings, and associative arrays whose subscripts can be numbers or arbitrary strings of characters. Variables are initialized to zero and an empty string, so there is usually no need to set initial values.

Awk automatically reads each line of each input file and splits each input line into fields, so one rarely needs code to explicitly read input or parse individual lines. There are also built-in variables that contain the number of the current input line and the number of fields on it, so those values do not have to be computed either. Such defaults eliminate boilerplate code and mean that many Awk programs are only a line or two long.

To illustrate, this program prefixes each line by its line number:

```
awk '{print NR, $0}'
```

NR is the number of the current input line, and $0 is the input line itself.

The following canonical example counts the number of occurrences of each word in its input, and prints the words and their counts at the end:

```
        { for (i = 1; i <= NF; i++) wd[$i]++ }
END { for (w in wd) print w, wd[w] }
```

The first line is an action with no pattern, so it is evaluated for each input line. The built-in variable NF is the number of fields on the current input line; this is computed automatically. The variable $i is the i-th field, again computed automatically. The statement wd[$i]++ uses that value, which is a word of the input, as a subscript in the array wd and increments that element of the array. The special pattern END matches after the last input line has been read. Note the two different kinds of for loop. The first is borrowed straight from C. The second loops over the elements of an array, so in this example, it prints a line for each distinct word of the original input, along with the count of how many times that word occurred.

Although Perl and later Python took over many potential applications, Awk is still widely used today; it's a core tool and there are at least four or five independent implementations, including Arnold Robbins's Gawk and Michael Brennan's Mawk. Awk certainly has some questionable design decisions and dark corners, but I think it gives the most bang for the programming buck of any language—one can learn much of it in 5 or 10 minutes, and typical programs are only a few lines long. It doesn't scale well to big programs, though that hasn't stopped people from writing Awk programs that are thousands of lines long.

Sed is popular as well, a frequent component of shell pipelines. I even have a bumper sticker that says

"Sed and Awk: together we can change everything."

It's worth noting that Sed, Awk, Make, Yacc and Lex all implement some flavor of the *pattern-action paradigm*. Programs in these languages consist of a sequence of patterns and actions; the basic operation is to check the input against each pattern and when a pattern is matched, perform the corresponding action. Patterns and actions may sometimes be omitted, in which case a default behavior takes place.

For example, grep, Sed and Awk can all be used to match a single regular expression. The following three commands are equivalent if the specific regular expression is valid for each:

```
grep re
sed -n /re/p
awk /re/
```

The pattern-action paradigm is a natural way to think about computations that are primarily a sequence of tests and actions.

Figure 5.15: Awk in popular culture

5.5 Other languages

The Unix programming environment, its language-development tools, a rich set of potential application domains, and of course local experts in everything from compilers to language theory and algorithms, led to the design and implementation of other languages as well. I'm not going to dive deeply into any of these, but it's worth a quick look at a partial list.

There's no need to understand any of the details in these examples. The real lesson is that wide-ranging interests, language expertise, and tools like

Yacc and Lex made it possible for members of the Center to create new languages for new application areas relatively easily. This would have been much harder without that combination of factors, and I think that many of these interesting languages would not have existed otherwise.

The most significant example is C++, which started in 1979 when Bjarne Stroustrup joined 1127, fresh from his PhD at Cambridge. Bjarne was interested in simulation and operating systems, but existing languages didn't really satisfy his needs. Accordingly, he took some of the good ideas from the closest match, Simula, and merged them with C. The result, which combined the ideas of object-oriented programming with the efficiency and expressiveness of C, was called "C with classes," and dates from 1980.

This proved to be a good combination, and the language prospered. In 1983, it acquired the name C++, Rick Mascitti's pun on the C ++ increment operator. Today C++ is one of the most widely used programming languages, at the heart of the implementation of Microsoft's Office suite, big parts of Google's infrastructure, your favorite browser (whatever it is), many video games, and much other behind-the-scenes software.

Bjarne was a member of my department for about 15 years, and as described earlier, he used to drop in frequently to talk out design decisions, so I got to see the evolution of C++ from the beginning. At least in the early days I understood it, but it's now a much bigger language and I'm barely literate in it.

C++ is often criticized for its size, and sometimes for some of the syntax that it inherited from C. I know from years of conversations that there isn't anything in the language for which Bjarne didn't have a good reason. It also was a sound engineering and marketing decision to make C++ a superset of C, even though that required including many of C's syntactic and semantic rough spots. If Bjarne had not aimed at C compatibility, C++ would have had much less chance of success. It's hard to establish a new language; making it compatible at both source level (for cultural familiarity) and object level (use of existing C libraries) was crucial, and at the time so was making it as efficient as C.

A handful of significant other languages, not yet mentioned, also came out of 1127.

Stu Feldman and Peter Weinberger wrote the first Fortran 77 compiler, called f77. As a language, Fortran 77 was somewhat better than the Fortran 66 that I had papered over with Ratfor, though it still didn't have a sensible set of control-flow statements. In any case, f77 was challenging to build but worthwhile, since it was heavily used by numerical analysts in 1127 on both the PDP-11 and the VAX.

In a related effort, Stu and Dave Gay wrote `f2c`, which translated Fortran into C; `f2c` made it possible to use Fortran on systems where no compiler was available, or where a Fortran compiler was an expensive commercial product.

Gerard Holzmann (Figure 5.16), who joined 1127 from the University of Technology in Delft, had always been interested in photography. In the early 1980s he conceived the idea of a programming language for making algorithmic transformations of digital image files. He called it Pico:

"Originally the name indicated its size, later it was more easily understood as an abbreviation of 'picture composition.' "

Pico is another example of a pattern-action language. It defines new images by evaluating a user-defined expression once for every pixel in the original image; expressions can refer to values, coordinates, various functions, and parts of other images. Such expressions can lead to entertaining transformations, many of which appeared in *Beyond Photography*, a book that Gerard published in 1988 to describe and illustrate Pico. (Figure 5.17 is an example.) Not surprisingly, Pico was implemented in C with a Yacc parser.

Figure 5.16:   Gerard Holzmann, ~1984 (Courtesy of Gerard Holzmann)

Gerard also created another specialized language-based tool, Spin, for analyzing and checking the correctness of software systems that involve

Figure 5.17: Gerard Holzmann, transformed by Pico

separate communicating processes. Spin can be used to verify that a particular system is logically correct, free of defects like deadlock where no progress can be made. ("After you." "No, after you.") Spin is an excellent example of research in 1127 into how to represent the interactions between separate processes over time, along with some first-rate software engineering to make a system that's easy to use and that runs fast enough to be useful. Spin models are written in another special-purpose language called Promela (protocol metalanguage), also implemented with Yacc.

Spin is thriving, with many thousands of installations and an annual conference. It has been used to verify a large number of systems, ranging from hardware designs to railway signaling protocols.

Bob Fourer, Dave Gay and I designed and implemented AMPL, a language for specifying optimization problems like linear programming. Bob was a professor of management science and operations research at Northwestern University, and had long been interested in helping people create mathematical optimization models. Our work on AMPL began when he spent a sabbatical at the Labs in 1984.

The AMPL language makes it easy to define models that describe particular optimization problems, like finding the best way to ship goods from factories to stores, given data for shipping costs, expected sales at each store, the manufacturing capacity of each factory, and so on. Optimization problems are written in an algebraic notation for systems of constraints that have to be satisfied, and an objective function to be maximized or minimized.

Optimization problems like these are at the heart of many industries: airline crew scheduling, manufacturing, shipping and distribution, inventory control, advertising campaigns, and an enormous variety of other applications.

I wrote the initial AMPL implementation in C++, along with a Yacc grammar and (I think) Lex for lexical analysis. It was my first serious C++ program, but I soon relinquished the code to Dave Gay.

AMPL is perhaps the only widely used language originating in 1127 that is proprietary. (The language itself can't be copyrighted, but there are at this point no open-source implementations that I am aware of.) AT&T started licensing AMPL to companies a few years after it was first created. When Dave and I retired from Bell Labs, the three of us formed a small company, AMPL Optimization, to continue AMPL development and marketing. Eventually we bought the rights from Bell Labs so we could go our own way. The company remains small but is a significant player in its niche market.

In the early 1980s, Rob Pike (Figure 5.18) and Luca Cardelli experimented with languages for concurrency, especially for interactions with input devices like mice and keyboards; that led to the names Squeak and Newsqueak. The ideas from Newsqueak eventually found their way into the concurrent languages Limbo and Alef that were used in Plan 9, and a decade later into the Go programming language, which was created at Google in 2008 by Rob Pike, Ken Thompson and Robert Griesemer.

5.6 Other contributions

Most of the emphasis so far has been on system software, especially languages, since that's what I know best, but I should mention some significant activities in scientific computing, communications, security and hardware, since they were often influential, and of course they all had substantial software components. These don't all fall neatly into the 7th edition time frame.

Scientific computing

As might be expected for a scientific research operation, Bell Labs was involved very early in the use of computers for modeling and simulation of physical systems and processing, a natural extension of mathematical research. It was also a validation of Dick Hamming's prediction that computing would displace laboratories. The focus was on numerical linear algebra, differential and integral equations, function approximation, and mathematical libraries that could make the best known methods widely available.

Figure 5.18: Rob Pike, ~1984 (Courtesy of Gerard Holzmann)

Phyllis Fox was a pioneer in this kind of numerical computation, and was a major contributor to the PORT library for Fortran programmers. PORT defined machine-specific constants for values like the ranges of numbers that would be different for different computers; its libraries assured portability of Fortran code across different computers.

The PORT library was a large body of work, ultimately with 130,000 lines of Fortran in 1,500 programs and a lot of documentation. Barbara Ryder and Stu Feldman developed a Fortran compiler called PFORT, which checked that Fortran code was written in a portable subset of standard Fortran. Norm Schryer wrote a program to check the arithmetic operations of computers, which often differed greatly in how they did floating-point arithmetic. This was especially important since it predated the development of standards for floating-point behavior.

Eric Grosse and Bill Coughran developed algorithms for semiconductor modeling and simulation, circuit analysis, and visualization, especially for semiconductor design and fabrication. Much of the numerical software developed at Bell Labs was distributed worldwide through the Netlib repository of mathematical software, which is still widely used by the scientific computing community. Other numerical analysts who made significant contributions to Netlib and the larger community included Dave Gay, Linda Kaufman, and Margaret Wright.

AT&T's 800-number directory

Eric Grosse's experience with software distribution was a big help in a fun unrelated project: in 1994, Eric, Lorinda Cherry and I put AT&T's 800-number directory on the then brand new Internet. The intent was for AT&T to gain some experience with providing a real Internet service (and with the Internet itself), perhaps to generate extra calls to 800 numbers, and even ultimately to provide revenue through enhanced services like display advertisements. In addition, we hoped for some modest public relations benefit from providing something of real value, not merely a teaser like so many Internet offerings at the time.

We obtained a database snapshot of 157,000 records, about 22 MB, in August 1994 and in a few hours had it running on a local computer as a searchable and browsable web site. Convincing AT&T management to make it public was a much slower process. After much internal deliberation, however, managerial reluctance and inertia eventually succumbed to a rumor that AT&T's competitor MCI was about to offer some Internet service, and the directory went public on October 19, 1994. It was AT&T's first web service.

The political delay was mildly frustrating, but the data story was instructive. The database was replete with errors, a host of listings that clearly no one had ever looked at with a critical eye, such as 9 different spellings of Cincinnati. (They would make a good example of a regular expression.)

In spite of its imperfections, the directory service garnered some public relations benefit: it was listed for a while as the first of the "Cool Links" on the Yahoo WWW Guide. (Yahoo itself was founded early in 1994, and its indexing was entirely manual.) AT&T scored a minor victory over MCI by being first with a useful service, though it was a near thing. The directory did make AT&T appear to be involved in the Internet and it stimulated a significant amount of internal discussion and plans to make plans for planning further services. If nothing else, the experience brought home the remarkable rate at which the Internet was growing and changing; as my unofficial report said at the time, "We have to learn to act quickly in this area if we are to be in it at all."

UUCP

In the mid 1970s, Mike Lesk wrote UUCP, the Unix to Unix copy program. It was used to send files from one Unix system to another, generally using ordinary phone lines. Those were slow and sometimes expensive, but they were ubiquitous and most Unix systems of the time had some kind of dial-up access, though fewer had dial-out capabilities, since that would involve paying for phone calls.

Although UUCP was primarily used for software distribution, it was also the basis of remote command execution, mail transfer and news groups long before the Internet was widely available. Usenet, which was one of the first worldwide information distribution systems, was based on UUCP.

The first UUCP distribution was included in the 7th edition; it was refined, ported to other operating systems, and made open source over the next few years. UUCP usage finally died out as the Internet became the standard communications network and its protocols took over.

Security

People in the Unix community had been concerned with security from the earliest days, an interest that came in part from Multics, and in part from experience in cryptography.

One security concern was to allow file system users to control access to their files. The Unix file system used 9 bits for each file to control what kinds of access could be made to the file. The file's owner had three bits, independently controlling read access, write access, and execute access. For a regular text file, read and write would normally be allowed for the owner, and execute would not, unless the file was an executable program or a shell script. There were three more bits for the owner's group, which might be something like a department or a project or a faculty/student distinction. The final three bits applied to all other users.

This mechanism was substantially simpler than what Multics had provided, but it has served well for a long time. For example, standard commands like editors, compilers, shells, and so on, are owned by a privileged account, typically the root user, which can read, write and execute them at will, but ordinary users can only execute (and perhaps read) but not write them. Note that it is possible to execute a program without being able to read its contents; in that way, the program can safely contain protected information.

One early refinement was a 10th permission bit, called the *setuid* (set user id) bit, for each file. When this bit is set, and the file is executed as a program, the user id for checking permissions is not the person running the program but the owner of the file. If the bit is set, an ordinary user can run a program with the privileges of the program's owner. This is used for programs that manipulate the file system to create directories, rename files, and so on: programs that execute privileged system calls are owned by a super-user that has unrestricted access. Naturally, setuid programs have to be carefully written and managed; if not, they can open large holes in system security. Setuid was invented by Dennis Ritchie, who received a patent for it in 1979.

As mentioned earlier, the idea of passwords originated with CTSS and was followed through in Multics and then into Unix. A text file called /etc/passwd contains one line for each user, with login name, user id number, password, and a few other fields. From earliest times, the password was stored in a hashed form, not in the clear. Hashing is a form of scrambling for which the only practical way to recreate the original is to try all possible passwords. This meant that anyone can read the password file but can't use the hashed passwords to login as someone else.

That's the theory. If the hashing isn't good or if people choose passwords poorly, however, decryption may be possible. Ken and Bob Morris collected password files from a variety of Unix systems and experimented with dictionary attacks, trying plausible passwords to see if they hashed to what was stored in the password file. Their studies in the mid 1970s showed that 10 to 30 percent of passwords could be obtained this way.

Dictionary attacks are still effective, though the technology on all sides is more sophisticated. One would hope that today's users are more aware of the perils of weak passwords, but recent lists of frequently used passwords suggest that they are not. (As an aside, this attack was also used in the Morris Worm of 1988, when Robert T. Morris, son of Bob, inadvertently released a program that tried to log in to Unix systems on the Internet and propagate itself. One of its mechanisms was to use a dictionary of likely passwords, like "password" and "12345.")

Bob wrote the original Unix `crypt` command. He retired from Bell Labs in 1986, to become the chief scientist at the National Security Agency (NSA), which suggests that he did know a fair amount about computer security and cryptography. He died in 2011 at the age of 78.

Cryptography was an on-going interest of several 1127 members, including Bob, Ken, Dennis, Peter Weinberger and Fred Grampp (Figure 5.19). (Dennis's web page tells some interesting behind-the-scenes stories.) Although encryption today is all done with software, it was done by mechanical devices during the Second World War. The German military used the Enigma machine, and somehow Fred obtained one. One story has it that he bought it on the open market; another says that his father, an American GI, brought it back from Germany after the end of the war. Fred kept it at Bell Labs and when he died left it to Ken Thompson, where it sat in the bottom drawer of Ken's file cabinet, across the hall from me.

Figure 5.19: Fred Grampp, ~1984 (Courtesy of Gerard Holzmann)

One day I borrowed it for a class I was teaching at Princeton that included a lecture on cryptography. I asked if anyone had ever seen an Enigma. No, no one had, so I pulled it out from under the table. I have never before or since seen such interest from my students, some of whom were literally standing on the table to get a proper look. Ken subsequently donated the Enigma to a museum.

When Ken and Dennis won the Turing Award in 1983, Ken's prescient talk, "Reflections on Trusting Trust," explained a series of modifications that he could make to a compiler that would eventually install a Trojan horse in the login program for a system.

> "You can't trust code that you did not totally create yourself. (Especially code from companies that employ people like me.) No amount of source-level verification or scrutiny will protect you from using untrusted code."

As he noted, the same kinds of tricks can be applied to hardware, where they will be even harder to discover. Things have not gotten better in the interim, and the paper is still highly relevant today.

Hardware

Software was the primary activity in 1127, but hardware interests were well represented too. In the early days, it often required hardware expertise to connect odd devices to the PDP-11; examples included the Votrax voice synthesizer, telephone equipment, typesetters, and a variety of network devices. This led over the years to the development of a suite of computer-aided design (CAD) tools, and involved a number of people, like Joe Condon, Lee McMahon, Bart Locanthi, Sandy Fraser, Andrew Hume, and others that I am surely forgetting.

Bart used the CAD tools to design and build a bitmap terminal in the early 1980s. At the time most terminals could only display 24 rows of 80 fixed width and fixed height ASCII characters. By contrast, a bitmap terminal displayed a large array of pixels, each of which could be independently set to a value, like the screens of all laptops and cell phones today, though the first bitmap displays were monochrome. Bart originally called his bitmap terminal the Jerq, an oblique reference to a somewhat analogous device called Perq offered by a Pittsburgh company named Three Rivers.

The Jerq started life with a Motorola 68000 processor, which was popular at the time (it was often used in workstations, for example), but both the name and the implementation fell victim to corporate politics. The Jerq was renamed the Blit (after the bitblit operation for rapidly updating screen contents) but was made only in small quantities. Western Electric, still the manufacturing arm of AT&T, redesigned it to use a Bellmac 32000, a processor chip designed by Bell Labs and manufactured by Western Electric. "Blit" was replaced by the catchy "DMD-5620." The redesign cost a full year and AT&T lost whatever chance it might have had to compete in the growing workstation marketplace.

Rob Pike wrote most of the operating system for the Blit and the 5620. Its most novel aspect was that computation could proceed in multiple overlapping windows. Overlapping windows had been seen before, but only one of them could be active at a time. Rob got a patent on the improvement.

The 5620 was a good graphics terminal, though physically heavy and bulky. I used it to write graphical programs such as a Troff previewer. It was also the environment in which Rob Pike wrote a series of mouse-based text editors, one of which I use by preference even today; this book was written with Sam.

There was also a sustained interest in integrated circuits and VLSI (Very Large Scale Integration). In 1980, Center 1127 offered a three-week crash course in integrated circuit design taught by Carver Mead of Caltech. Lynn Conway and Carver had literally written the book (*Introduction to VLSI Systems*, 1980) on how to design and implement integrated circuit chips, and their course had already been given at a number of universities. Carver had a gift for telling a carefully calibrated sequence of lies about how circuits worked. The simplest version was that when a red line crossed over a green line, that made a transistor. Of course this gross over-simplification was unmasked, to be replaced by another more sophisticated lie, which in turn would be further refined.

As a result of this excellent instruction, everyone in Carver's class was able to design and build some kind of experimental chip after a couple of weeks of training. The chips were fabricated at the Bell Labs plant in Allentown, Pennsylvania, and returned for experimentation. At the time, Bell Labs was using state of the art 3.5 micron technology; today's circuits are usually 7 to 10 nanometers, an improvement of at least 300 in line widths, and thus about 100,000 in the number of devices in a given area.

My chip, a simple chess clock, never worked, thanks to a grievous logic error that was obvious in retrospect. Several people developed support tools as well as their own chips. My contribution was a program to help route on-chip wiring, so it was a productive few weeks in spite of my failure as a chip designer.

Over the years, at least half a dozen people in my department did VLSI in one form or another—algorithms for checking layouts, simulators, reverse-engineering, and some theoretical research. I could keep up with what they were doing, barely, thanks to Carver's course.

The Center's interest in VLSI persisted for a long time, ultimately leading to Dave Ditzel and Rae McLellan's CRISP (C Reduced Instruction Set Processor) microprocessor, which was one of the earliest RISC processors. "RISC" is an acronym for Reduced Instruction Set Computer, a way to design processor architectures that are simpler and more regular than ones like the VAX-11/780.

CRISP aimed at an instruction set that would be well suited for the output of a C compiler. To design it, Dave worked closely with Steve Johnson. After discussing potential architecture features, Steve would modify the Portable C compiler and run benchmarks to see what effect these features would have on performance, a great example of hardware/software co-design.

AT&T eventually sold a version of CRISP under the name Hobbit. It was intended for the Apple Newton, one of the first personal digital assistants, but neither Newton or Hobbit was a commercial success. Dave left Bell Labs for Sun Microsystems, and in 1995 founded Transmeta, which focused on low-power processors.

Even though CRISP itself did not succeed commercially, Unix and C had a large impact on computing hardware in the 1980s and 1990s. Most successful instruction set architectures were well-matched to C and Unix. Not only did Unix and C portability enable universities and especially companies to create new architectures and rapidly port software, but it had the effect of requiring that the instruction set was good for C code, while tending to eliminate features that were hard to compile from C programs. CPU design methodology like that used by Johnson and Ditzel used statistical analysis of programs, so C code statistics tended to favor things that made C fast. Unix and C had a critical mass that caused CPU design in the 1980s and 1990s to orbit around them; nobody was successfully building CPUs tuned for other languages.

Chapter 6

Beyond Research

> "The Unix operating system presently is used at 1400 universities and colleges around the world. It is the basis for 70 computer lines covering the microcomputer to supercomputer spectrum; there are on the order of 100,000 Unix systems now in operation, and approximately 100 companies are developing applications based on it."
> R. L. Martin, *Unix System Readings and Applications*, Vol. 2, 1984

After a few years in the lab in Center 1127, Unix began to spread, both inside Bell Labs and outside, the latter primarily through universities, which were able to get the source code for the whole system for a nominal "media fee" under a trade secret agreement. This was definitely not "open source": the system could be used only for educational purposes, and licensees could talk only to other licensed users about their experience and what they were doing with Unix. Nevertheless, the community grew rapidly, user groups sprang up all over the world, and major technical innovations took place, for example ports of the system to different kinds of hardware and new mechanisms for accessing the Internet.

Bear in mind that many of the activities described in this chapter were going on in parallel with, or even years before, those of the previous chapter. This can make the chronology a bit confusing.

6.1 Programmer's Workbench

The patent department at Bell Labs was the first "customer" outside of the research area, but other groups found Unix useful as well, and the system started early to spread within Bell Labs development groups and other parts of AT&T.

With over a million employees, AT&T was a very big company, with many computer systems for managing the data and operations that supported telephone services. These systems provided technician interfaces and support for AT&T, keeping track of equipment and customers, monitoring systems in the field, logging events, trouble-shooting, and the like. Collectively, these systems were called "operations support systems."

One of the first major Unix installations outside of Research was a group based in Piscataway, New Jersey, about 15 miles from Murray Hill, that in 1973 began to develop tools for programmers who did software development for large-scale production environments. The collection became known as the Programmer's Workbench or PWB.

Most operations support systems at AT&T ran on large mainframe computers from IBM and Univac that had their own proprietary operating systems, such as IBM's OS/360. PWB provided facilities for creating and managing the software that ran on such computers. In effect, PWB Unix served as a uniform front end for a diverse set of large non-Unix computer systems; the mainframes were treated like peripheral devices.

One major PWB service was remote job entry, a set of commands for sending jobs to target systems and returning the results, including job queueing, status reports, notifications, logging and error recovery. Remote job entry fit well with the Unix approach of using small tools that could be connected in a variety of ways, and then encapsulated in shell scripts for convenient use by non-experts.

To support this kind of programming, John Mashey (Figure 6.1) enhanced the 6th edition shell to create the PWB shell, which provided much better programmability, including a general *if-then-else* for decision-making, a *while* for looping, and shell variables for storing text. He also invented a search path mechanism, so that by setting a particular shell variable, any user could specify a sequence of directories to be searched for commands. The search path made it easy for groups of users to collect programs in project directories, rather than having to install commands in system directories for which they might not even have permission. As John said,

> "We had a large (1,000+) population of users who were not usually C programmers, who worked in groups in shared environments. They wanted to share their own sets of commands by lab, department, group. They often shared systems with others and none could be super-users."

John also added a mechanism so that if a file was marked executable, it would either be executed as a regular command, or passed to a shell if it was a script. All of these features were in place by early 1975, and refined over the next year as more and more people began using the PWB shell. John's paper "Using a Command Language as a High-Level Programming Language" reported on experience with over 1,700 shell procedures:

> "By utilizing the shell as a programming language, PWB users have been able to eliminate a great deal of the programming drudgery that often accompanies a large project. Many manual procedures have been quickly, cheaply, and conveniently automated. Because it is so easy to create and use shell procedures, each separate project has tended to customize the general PWB environment into one tailored to its own requirements, organizational structure, and terminology."

As noted in the previous chapter, John's improvements soon found their way back into the shell that Steve Bourne wrote.

Figure 6.1: John Mashey, ~2011 (Courtesy of Twitter)

Another important PWB product was the Source Code Control System (SCCS), which was written in 1972 by Marc Rochkind. SCCS was the first of a sequence of programs for managing large code bases that were being worked on by several programmers at the same time.

The basic idea of SCCS was to let programmers check out part of the code base to work on; this locked that part so other programmers couldn't change it until the lock holder unlocked it. This prevented multiple programmers from making inconsistent changes to the same piece of code at the same time. Of course there were still problems; for example carelessness or crashes could leave code locked even though no one was working on it, and if the locked regions were too big, that slowed down simultaneous changes. But the idea was crucial for software development that involved multiple people working on the same code base, and it's even more important today with larger code bases spread across much larger developer communities that are more geographically dispersed. There's a clear evolutionary path from SCCS through RCS, CVS and Subversion, to Git, today's default standard version-control system.

Marc Rochkind also invented a tool that converted a set of regular expressions into a C program that would scan logs for occurrences of the patterns and print messages when one was found. This was such a neat idea that Al Aho, Peter Weinberger and I ~~stole~~ adapted and generalized it for the pattern-action model used in Awk.

PWB also included a collection of tools called the Writer's Workbench (WWB) that tried to help people to write better. John Mashey and Dale Smith, with encouragement from Ted Dolotta, created a set of generic Troff commands, the Memorandum Macro or mm package, that was widely used both inside AT&T and outside for producing documentation.

In addition, WWB offered a spell checker and programs for finding punctuation mistakes, split infinitives, and double words (often the result of errors made while editing). There were tools for checking grammar and style, and for assessing readability. The core component was a program called *parts*, by Lorinda Cherry. It assigned parts of speech to the words of a text, and although imperfect, it yielded statistics on the frequency of adjectives, compound sentences, and the like. WWB was developed in the late 1970s, just about the time when computers were being more frequently used by writers, and WWB got some good press, including an appearance on national TV on NBC's *Today* show for two of its creators, Lorinda and Nina McDonald.

As an example of how computing hardware has become cheaper and more powerful over the years, a 1978 PWB paper by Ted Dolotta and Mashey described the development environment, which supported over a thousand users: "By most measures, it is the largest known Unix installation in the world." It ran on a network of 7 PDP-11's with a *total* of 3.3 megabytes of primary memory and 2 gigabytes of disk. That's about one thousandth of a typical laptop of today. Would your laptop support a population of a million users?

6.2 University licenses

In 1973, AT&T began licensing Unix to universities for a nominal fee, though most licenses were for the 6th edition, which became available in 1975. There were some commercial 6th edition licenses as well, but these cost $20,000, which would be more like $100,000 today. Licenses did include all the source code, but came with no support whatsoever.

One of the most active license recipients was the University of California at Berkeley (UCB), where a number of graduate students made major contributions to the system that eventually became the Berkeley Software Distribution (BSD), one of two main branches that evolved from the original Research Unix.

Ken Thompson spent a sabbatical year at Berkeley in 1975 and 1976, where he taught courses on operating systems. Bill Joy (Figure 6.2), a grad student at the time, modified the local version of Unix, and added some programs of his own, including the `vi` text editor, which is still one of the most popular Unix editors, and the C shell `csh`. Bill later designed the TCP/IP networking interface for Unix that is still used today. His `socket` interface made it possible to read and write network connections with the same `read` and `write` system calls as were used for file and device I/O, so it was easy to add networking functionality.

Bill occasionally visited Bell Labs during the mid to late 1970s. I remember one evening when he showed me the new text editor that he was working on. By this time, video display terminals had replaced paper terminals like Teletypes, and they enabled a much more interactive style of editing.

In `ed` and other editors of the time, one typed commands that modified the text being edited, but they did not continuously display the text; instead, if an editing command changed some text, it was necessary to explicitly print the new line. In `ed`, one might say

```
s/this/that/p
```

to substitute `this` into `that` in the current line and print the result. Other commands made it possible to change occurrences in multiple lines, search for lines, display ranges of lines, and so on. In the hands of experts, `ed` was very efficient, but not intuitive for newcomers.

Bill's editor used cursor addressing to update the screen as text was being edited. This was a major change from the line-at-a-time model: one moved the cursor to `this` (perhaps by using a regular expression), typed a command like `cw` ("change word"), and then typed `that`, which immediately replaced the original.

I don't recall what I said at the time about the editor itself (though today `vi` is one of the two editors that I use most often), but I do remember telling Bill that he should stop fooling around with editors and finish his PhD. Fortunately for him and for many others, he ignored my advice. A few years later, he dropped out of graduate school to co-found Sun Microsystems, one of the first workstation companies, with software based on Berkeley Unix, including his fundamental work on the system, networking and tools (and his `vi` editor). I often cite this story when students ask me for career advice—older is not always wiser.

Figure 6.2: Bill Joy, ~1980 (Courtesy of Bill Joy)

6.3 User groups and Usenix

Since AT&T offered no support at all to Unix licensees, users were forced to band together to help each other, in what eventually became regular meetings with technical presentations, software exchanges, and of course social activities. This idea was certainly not original with Unix; the SHARE user group for IBM systems was established in 1955 and is still active, and there were user groups for other hardware manufacturers as well.

The first Unix user group meeting was held in New York in 1974, and user groups gradually sprang up all over the world. In 1979, Ken and I attended the first meeting of the UKUUG, which was held at the University of Kent in Canterbury. It was quite an experience for me, the first time I had visited England. Ken and I flew to Gatwick Airport on Laker Airways, the first of the cheap trans-Atlantic airlines. We drove to Salisbury and visited the cathedral and Stonehenge, then went on to Canterbury for the meeting (and to visit the cathedral). Afterwards I spent a few days in London as a wide-eyed tourist.

I have since visited several other countries, using Unix user group meetings as an excuse and a great way to meet some very nice people. The most memorable was a trip to Australia in 1984, again with Ken, where the meeting was held in the Sydney Opera House. I gave a talk on the morning of the first day, then spent the rest of the week watching the harbor from a window of the conference room; it was so fascinating that I have no memory of any of the other talks.

The user groups evolved into an umbrella organization called "Unix User Groups," which was renamed USENIX (Usenix from now on) after AT&T complained about mis-use of the Unix trademark. Usenix now runs an extensive series of professional conferences and publishes a technical journal called ";login:". Usenix played a significant role in spreading Unix, with conference presentations and tutorials on many subjects. It also distributed UUCP and ran the Usenet news system.

6.4 John Lions' Commentary

John Lions (Figure 6.3), a professor of computer science at the University of New South Wales in Sydney, was an enthusiastic early adopter of Unix, and used it extensively in courses on operating systems as well as for general educational and research support at UNSW.

In 1977, John wrote a line-by-line "commentary" on the 6th edition source code. Every part of the source code was explained in detail, so one could see how it worked, why it was the way it was, and how it might be done differently. John also produced a number of excellent students, several of whom wound up at Bell Labs.

In the original printing, the *Commentary* was in two separate volumes, the code in one volume and the exposition in the other, so that they could be read side by side, though the authorized version (Figure 6.4), which finally

Figure 6.3: John Lions (Courtesy of UNSW)

appeared in 1996, is a single volume.

Although it could be shared among Unix licensees, the book was technically a trade secret since it contained AT&T's proprietary source code. Copies were carefully controlled, at least in theory; I still have my numbered copy (#135). But it was widely copied in the early days; the samizdat imagery on the cover in Figure 6.4 suggests that such copies were made clandestinely. Years later, reality was acknowledged and John's book was published commercially.

John spent a sabbatical year in 1978 at Murray Hill, across the hall from me in the office that later belonged to Dennis Ritchie. John died in 1998 at the age of 62. His contributions have been remembered in a chair of computer science at UNSW. The chair was funded by donations from alumni and friends, including Ted Dolotta's 1998 auction of his California UNIX license plate, bought by John Mashey.

One comment in the Unix source has become famous, thanks to the Commentary. Line 2238 says

```
/* You are not expected to understand this. */
```

CHAPTER 6: BEYOND RESEARCH

Figure 6.4: John Lions' Commentary on 6th Edition Unix

As mentioned earlier, Dennis died in October 2011. I used this comment as the core of a tribute to Dennis during a memorial gathering at Bell Labs the following year.

The Unix kernel code was jointly written by Dennis and Ken Thompson. As far as I know, Ken always subscribed fully to the idea that good code doesn't need many comments, so by extrapolation, great code needs none at all; I think that most of the comments in the kernel come from Dennis. The specific comment, which you can find on line 2238, is famous for its dry wit, and it was widely available on t-shirts and the like for years. As Dennis himself said,

> "It's often quoted as a slur on the quantity or quality of the comments in the Bell Labs research releases of Unix. Not an unfair observation in general, I fear, but in this case unjustified."

If you go back and look, you can see that the comment comes just after a much longer comment that describes the context-switching mechanism that swaps control between two processes, and it really was trying to explain something difficult. Dennis went on to say,

> " 'You are not expected to understand this' was intended as a remark in

the spirit of 'This won't be on the exam,' rather than as an impudent challenge."

I mentioned earlier that Nroff and Troff were difficult tools to master. The final paragraph of the acknowledgments section of the Lions *Commentary* suggests that John would concur:

> "The cooperation of the nroff program must also be mentioned. Without it, these notes could never have been produced in this form. However it has yielded some of its more enigmatic secrets so reluctantly, that the author's gratitude is indeed mixed. Certainly nroff itself must provide a fertile field for future practitioners of the program documenter's art."

6.5 Portability

The 6th edition of Unix was mostly written in C, with a limited assembly language assist to access machine features that were otherwise not available, for example, setting up registers, memory mapping, and the like. At the same time, Steve Johnson had created a version of the C compiler that was "portable" in the sense that it could be straightforwardly retargeted to generate assembly language for architectures other than the PDP-11. This made it possible to move C programs to other kinds of computers merely by recompiling them. The most interesting program to port would obviously be the operating system itself. Would that be feasible?

The first port of Unix was done at the University of Wollongong in New South Wales, Australia, by Richard Miller; the target computer was an Interdata 7/32. Miller didn't use the portable C compiler. Instead he bootstrapped his way onto the Interdata by modifying the code generator of Dennis Ritchie's original C compiler. His version of Unix was working and self-sustaining by April 1977.

Independently, and without knowing of Miller's work, Steve Johnson and Dennis Ritchie ported Unix to a similar machine, the Interdata 8/32. Their goal was somewhat different, a version of Unix that was more portable, rather than a single port of the original as it was. They got their version running late in 1977. Steve Johnson recalls some of the background:

> "There was another pressure to make Unix portable. A number of DEC's competitors were beginning to grumble that regulated AT&T had too cozy a relationship with DEC. We pointed out that there were no other machines like the PDP-11 on the market, but this argument

was getting weaker. Dennis hooked me into the portability effort with one sentence: 'I think that it would be easier to move Unix to another piece of hardware than to rewrite an application to run under a different operating system.' I was all in from that point on."

Portability was a great step forward. Up to this point, operating systems had mostly been written in assembly language, and even if written in a high-level language were more or less tied to a particular architecture. But after projects like those done by Miller, and Johnson and Ritchie, porting Unix to other kinds of computers, though not trivial, was basically straightforward. This had major implications for the emerging workstation marketplace, where companies both old and new were building computers that were smaller and cheaper than minicomputers like the PDP-11 and the Interdata, and used different processors.

Workstations were meant to be personal machines for scientists and engineers, giving them a powerful and normally unshared computing environment. There were many examples of workstations, of which the ones from Sun Microsystems were the most commercially successful. Other manufacturers included Silicon Graphics, DEC, Hewlett-Packard, NeXT and even IBM. The first workstations, in the early 1980s, aimed for a megabyte of primary memory, a megapixel display, and a speed of one megaflop (a million floating-point operations per second), a so-called "3M" machine. For comparison, my elderly Macbook has 8 gigabytes of memory and a speed of at least a gigaflop; its display is little more than a megapixel, but the pixels are 24-bit color, not monochrome.

The workstation marketplace arose because technological improvements made it possible to pack serious computing horsepower into a small physical package and sell it for a modest price. The complete system price could be reasonable in part because software, including the operating system, was already available. There was no need for new manufacturers to create a new operating system—it was enough to port Unix and its accompanying programs to whatever processor the computer used. The workstation market was thus helped significantly by the availability of Unix.

The IBM Personal Computer (PC) first appeared in 1981. The PC and its many clones were typically 5 to 10 times cheaper than a workstation. Though they were originally not at all competitive in performance, they gradually improved, and by the end of the 1990s were at least as good. Any distinction between workstation and PC eventually blurred. Today, depending on application areas, such machines most often run Microsoft Windows, macOS, or a Unix or Unix-like system.

Chapter 7

Commercialization

"As Unix spread throughout the academic world, businesses eventually became aware of Unix from their newly hired programmers who had used it in college."
    Lucent web site, 2002

"Unix and C are the ultimate computer viruses."
    Richard Gabriel, "Worse is Better," 1991

It had been argued that AT&T was prohibited from selling Unix commercially because as a regulated public monopoly, if it did so, it would be competing with other operating system vendors, using revenues from telephone services to cross-subsidize Unix development. The closest that AT&T came to making a real business was to license Unix to corporate customers for $20,000 (in contrast with the nominal media fee for educational institutions) but in limited quantities and without support. That was enough to head off regulatory scrutiny.

7.1 Divestiture

By 1980, AT&T's position as a monopoly, regulated or not, was coming under attack. The US Department of Justice had begun an antitrust lawsuit against AT&T in 1974, on the grounds that AT&T controlled not only telephone service for most of the country but also the equipment used by its telephone companies, and thus that AT&T controlled communications for the whole country. The DoJ proposal was that AT&T should be forced to divest its manufacturing arm, Western Electric, which made the equipment.

AT&T proposed instead to address the situation by splitting the company into one part (called AT&T) that would provide long distance telephony, and seven regional operating companies ("Baby Bells") that would provide local phone service in their respective geographic areas. AT&T would retain Western Electric but the operating companies would no longer be required to buy equipment from it. AT&T would also keep Bell Labs.

The consent decree with the Department of Justice, by which AT&T divested itself of the operating companies, was finalized early in 1982, and took effect January 1, 1984.

Divestiture was an enormous upheaval, which in the long run led to misfortune for AT&T, and 20 years of subsequent misjudgments and poor business choices made Bell Labs into a shadow of what it had been when its mission was clear and its funding was adequate and stable.

In 1984, part of Bell Labs was spun off into a research organization originally called Bellcore ("Bell Communications Research") that was to provide research services for the Baby Bells. The Bellcore split took a fair number of people from Bell Labs research, primarily in communications areas, but also some colleagues from 1127, including Mike Lesk and Stu Feldman. At some point, the Baby Bells decided that they didn't need the kind of research that Bellcore provided, and Bellcore was purchased by another company, SAIC, renamed Telcordia, and eventually wound up owned by the Swedish telecom company Ericsson.

1984 was also the year that "Bell Labs" became "AT&T Bell Laboratories," since as part of the consent decree, AT&T was not permitted to use the name "Bell" except in this special case and only if preceded by "AT&T." We were strongly encouraged to use the full name at all times.

7.2 USL and SVR4

After divestiture, AT&T's inability or at least reluctance to sell Unix gave way to a substantial commercial effort by a part of the company that was organizationally far removed from Research. It was also somewhat removed physically, located in a building near Summit, New Jersey; surrounded by busy highways, it was informally known as "Freeway Island." The organization was originally called the Unix Support Group (USG) and eventually became Unix System Laboratories, or USL. The first USG was created by Berk Tague in 1973, to focus on operations support systems. Over time, USG broadened its activities to include external sales and marketing.

There was definitely a market for Unix; one might even say that AT&T had inadvertently created the market by giving Unix away to university students, who then wanted it when they entered the real world and worked for companies that could afford to pay real money for it. Beginning in 1984 USL marketed Unix aggressively, and worked hard to make it a professional commercial product. The culmination was a version called System V Release 4, or SVR4. AT&T invested substantial resources to making this version a standard, with reference implementations and careful definitions for both source and object compatibility. I think that this attention to standards and interoperability was important.

The ins and outs of how SVR4 evolved and AT&T's interactions with cooperators and competitors over a decade are intricate and uninteresting. I'm not going to say anything more about them, since in a way it's all moot: the industry focus has shifted almost entirely to Linux. The Wikipedia article on System V summarizes the situation, probably accurately, like this:

> "Industry analysts generally characterize proprietary Unix as having entered a period of slow but permanent decline."

Of course the operative word is "proprietary"; open-source versions like the BSD family described in the next chapter are alive and well.

AT&T's product line included the operating system and a variety of supporting software, including compilers for C, C++, Fortran, Ada and even Pascal, mostly based on Steve Johnson's portable C compiler. There was also a major standardization effort to ensure compatibility for source code and for binary formats in libraries.

At this point I was Bjarne Stroustrup's department head, which meant frequent interactions with USL about the evolution of C++. For the most part, these were amicable, but there were occasions when the different priorities of research and product management were visible. One heated discussion with a USL manager in 1988 went something like this:

> Manager: "You have to fix all the bugs in the C++ compiler, but you can't change the behavior in any way."
>
> Me: "That's not possible. By definition, if you fix a bug, the behavior is necessarily different."
>
> Manager: "Brian, you don't understand. You have to fix the bugs but the compiler's behavior can't change."

Pedantically I was right, but practically I can see what the manager was getting at—too much or too rapid change is a serious problem for those who are doing software development with new languages and tools.

USL set up a subsidiary in Japan called Unix Pacific, where the manager was Larry Crume, a Bell Labs colleague from years earlier and well known to many in the research group. This led to technical collaborations, and I got to visit Japan twice on the company's money. On one of these trips, some kind of feel-good exchange with NTT, Japan's major telephone company, there was a strikingly clear view of the pecking order. The executive director got to play golf with his NTT counterpart. The center director played tennis with his counterpart. Lowly department heads like me were offered a shopping trip in Tokyo, which I declined with thanks.

Although AT&T's efforts to commercialize Unix were not always successful, Unix standardization was invaluable for the whole community. There were occasional tensions between Research and the USL effort, but for the most part, USL had a large group of talented colleagues who made significant contributions to Unix and related software systems.

7.3 UNIX™

Sometime early in the life of Unix, Bell Labs' legal guardians decided that the name Unix was a valuable trademark that had to be protected, which was certainly a correct business decision. They were trying to avoid having the name become a generic term that could be used by anyone, as had happened to words like aspirin (in the US, though not everywhere), escalator, zipper and (much more recently) App Store.

As a consequence, however, people inside Bell Labs were required to use the name correctly. In particular, it could not be used as a standalone noun ("Unix is an operating system"). It had to be identified as a trademark and also had to appear as an upper-case adjective in the phrase "the UNIX™ operating system," which led to awkward sentences like "The UNIX™ operating system is an operating system." Rob Pike and I had to fight this battle over the title of our 1984 book *The Unix Programming Environment*, which would otherwise have been something like *The UNIX™ Operating System Programming Environment*. The eventual compromise: there is no footnote or trademark indicator on the cover, but on the title page there is an almost invisible asterisk and a footnote.

The klunky phrasing was a pain, especially for people who took their writing seriously, so there were workarounds and occasional attempts to evade the issue. For instance, in the standard Troff macro package `ms`, Mike Lesk added a formatting command that expanded into "UNIX" wherever it was used, and automatically generated a footnote on the first page where it appeared (with the name in upper case, of course). Normally the footnote said

† UNIX is a trademark of Bell Laboratories.

but if the command was used with an additional undocumented parameter, it would instead print

† UNIX is a footnote of Bell Laboratories.

I don't think that anyone ever noticed when we occasionally used this Easter egg, but the code is still there in the standard macro package.

Other uses of the word Unix for goods and services had nothing to do with operating systems, like the pens in Figure 7.1, the bookcases in Figure 7.2, and the fire extinguisher in Figure 7.3. They all seemed to be from outside the United States and thus beyond US trademark laws; the bookcases date from 1941, before Ken and Dennis were born. One other charming example is Unix baby diapers from a company called Drypers, which came up with "Unix" as a contraction of "unisex."

Figure 7.1: Unix pens (Courtesy of Arnold Robbins)

7.4 Public relations

There was always a steady flow of visitors to Bell Labs, and from the mid 1970s until well into the 1980s, a Unix presentation was a frequent stop for tourists. A small group of visitors would sit in a conference room while some member of the Center gave them a quick overview of what Unix was all about and why it was important to AT&T and the world. Mike Lesk and I probably did more of these demos than everyone else put together, which may have reflected defects in our personalities, since we both complained but in fact enjoyed them.

The visitors themselves ranged from ordinary mortals to "distinguished," a synonym for "influential" or "important for AT&T to impress" or merely

Figure 7.2: Unix sectional bookcases, 1941 (Courtesy of Ian Utting)

Figure 7.3: A Unix fire extinguisher

"big name." For example, in 1980 I did a demo for Walter Annenberg, founder of *TV Guide* (which is where he made the money that perhaps helped him to become Ambassador to the Court of St James, though that role was over by the time I showed him the wonders of Unix). As a measure of his importance, he was accompanied by Bill Baker, the president of Bell Labs.

My personal shtick often involved a demo of pipes, showing how programs could be fluidly combined to do quicky *ad hoc* tasks. I used a shell script for finding potential spelling errors in a document, since it was a neat example of a long pipeline and helped to make the point about how existing programs could be combined in novel ways.

The `spell` script originated with Steve Johnson. The basic idea was to compare the words of a document to the words in a dictionary. Any word that occurred in the document but not in the dictionary was potentially a spelling mistake. The script looked approximately like this:

```
cat document |
tr A-Z a-z |         # convert to lower case
tr -d ',.:;()?!' |   # remove punctuation, ...
tr ' ' '\n' |        # split words into lines
sort |               # sort words of document
uniq |               # eliminate duplicates
comm -1 - dict       # print lines found in input but
                     #             not in dictionary
```

All these programs already existed; comm, the most unusual, was used for tasks like finding lines that were common to two sorted input files, or lines that were in one input or the other but not both. The "dictionary" was /usr/dict/web2, the words in Webster's Second International dictionary, one per line, which we saw earlier.

Figure 7.4: Unix building blocks, ~1980 (Courtesy of Bell Labs)

One day I was scheduled to do a demo for William Colby, who at the time was the director of the Central Intelligence Agency (CIA), and thus clearly an important person. He too would be accompanied by Bill Baker, who as head of the President's Foreign Intelligence Advisory Board had serious spook credentials of his own.

I wanted to make the point about how Unix made some kinds of programming easy, but the spell script was not very fast and I didn't want the demo to drag on. So I ran the script ahead of time, captured the output in a file, then wrote a new script that merely went to sleep for two seconds, then printed the results that had been computed the day before:

```
sleep 2
cat previously.computed.output
```

This bit of demo engineering worked well; if he understood it at all, Mr Colby must have thought that spell checking ran fast enough. Of course there's a lesson here for everyone who has ever sat through a demo: be wary of what you're seeing!

The PR operation also made promotional movies that talked about the wonders of Bell Labs, including a number that featured Unix. Thanks to YouTube, I can see old friends (and myself) when we were all younger and in several cases had more and darker hair.

There was even a brief flurry of print advertising for Unix. I think the children's blocks in the advertisement in Figure 7.4 were my idea, for better or worse. The background, too hard to see here, is a Troff document that I provided.

Chapter 8

Descendants

"... from so simple a beginning endless forms most beautiful and most wonderful have been, and are being, evolved."
　　　Charles Darwin, *The Origin of Species*, Chapter 14, 1859

Unix began in the Computing Science Research Center in 1969. There were internal versions like PWB that supported the Programmer's Workbench tools, of course, but starting in 1975, external versions appeared as well, originally based on the 6th edition, and then later on the 7th edition, which appeared in 1979.

The 7th edition was the last Research version of Unix to be released and widely used. Three more editions were developed and used internally (predictably called 8th, 9th, and 10th) but by the time the 10th edition was completed in late 1989, it was clear that the center(s) of gravity of Unix development had moved elsewhere.

Two threads evolved from the 7th edition, one from Berkeley that built on the work of Bill Joy and colleagues, and another from AT&T as it tried to build a money-making business out of its Unix expertise and ownership. The timeline in Figure 8.1 is an approximation and omits numerous systems; reality was more complicated, especially in how the versions interacted.

8.1 Berkeley Software Distribution

In 1978, DEC introduced a new computer, the VAX-11/780. The VAX was a 32-bit machine with substantially more memory and computing horsepower than the PDP-11, while remaining culturally compatible with it. A 16-bit computer uses 16 bits for a memory address, while a 32-bit computer

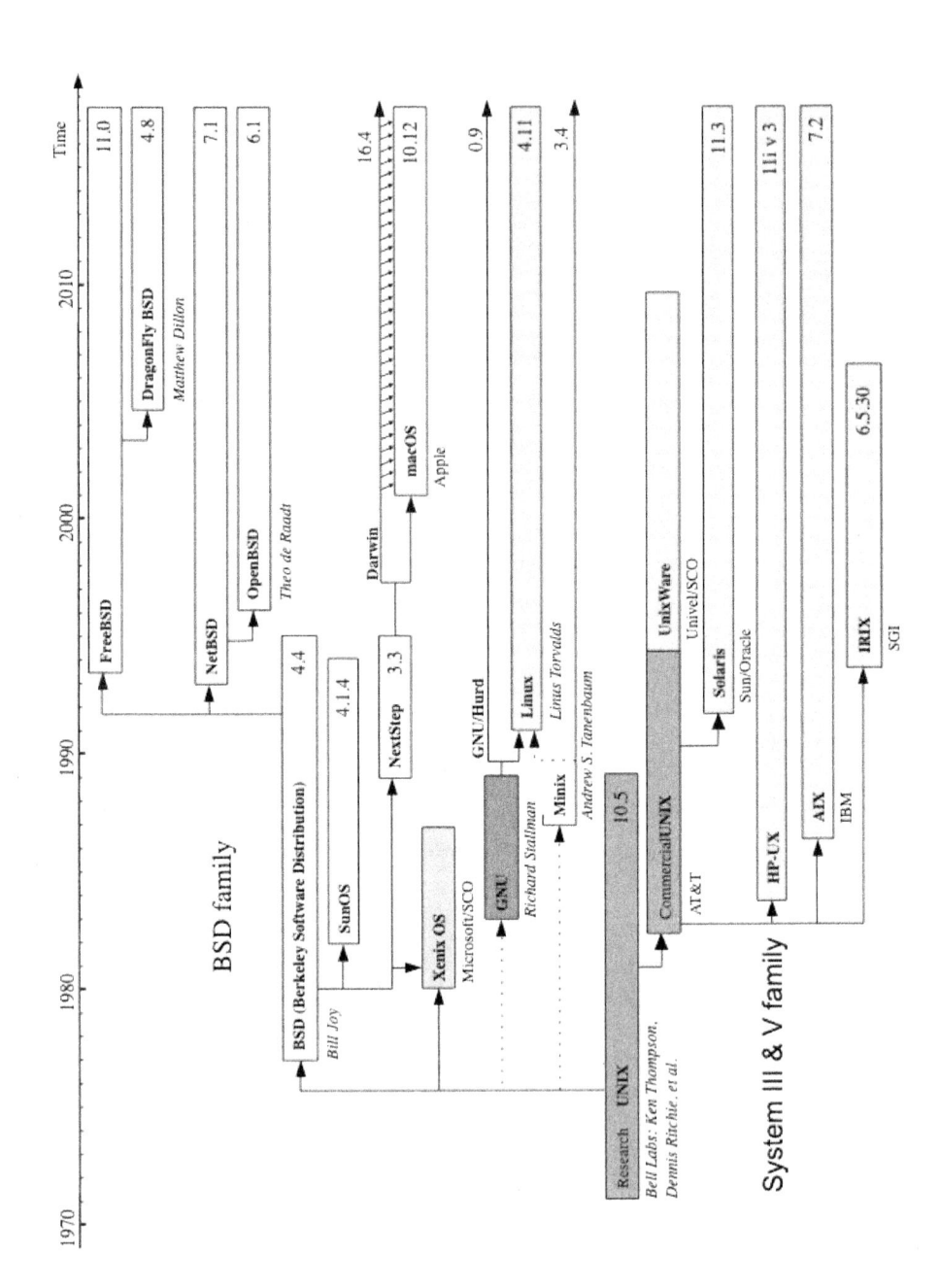

Figure 8.1: Unix timeline, from Wikipedia

uses 32 bits and thus can address a much larger amount of primary memory. When the VAX-11/780 first appeared, John Reiser and Tom London, who were in a research group at Bell Labs in Holmdel, New Jersey, ported the 7th edition to it, but their version, 32/V, did not use the virtual memory capabilities of the new machine and thus did not take full advantage of what the VAX could do.

Bill Joy and his colleagues at the Computer Systems Research Group at the University of California, Berkeley, started with Reiser and London's 32/V and added code to use virtual memory. This version quickly supplanted 32/V and the VAX itself became the primary Unix machine for most users as they outgrew the PDP-11. The Berkeley version was packaged and shipped to Unix licensees as BSD, the Berkeley Software Distribution. BSD descendants are still active, with variants like FreeBSD, OpenBSD and NetBSD all continuing development. NextSTEP, used for Apple's Darwin, the core of macOS, was also a BSD derivative.

One of the early Berkeley distributions formed the base of SunOS, which was used on computers from Sun Microsystems, co-founded by Bill Joy. Others spun off a few years later into the BSD variants mentioned above. All of these eventually were reimplementations that provided the same functionality but with entirely new code. Once rewritten, they were free of AT&T code and thus did not infringe AT&T's intellectual property.

Another spinoff was created for NeXT, which was founded by Steve Jobs in 1985. The NeXT workstation had a variety of innovative features, and was an early example of the elegant and polished industrial design that Apple users are familiar with. I was in the audience at Bell Labs on December 11, 1990, when Jobs gave a demonstration of the NeXT. It was a very nice machine, and it was the only time that I can recall thinking "I want one of those" about any technological gadget. I had obviously been seduced by the famous Jobs "reality distortion field." When he did another presentation at the Labs three years later, there was no such effect, and I don't even remember what he was showing off.

Although the NeXT computer itself was not a commercial success, the company was bought by Apple in 1997, and Jobs returned to his old company, becoming CEO within a year. One can still see some of the NextSTEP operating system legacy in names like `NSObject` and `NSString` in Objective-C programs.

The timeline reveals another little known fact: in the 1980s Microsoft distributed a version of Unix called Xenix; Figure 8.2 is part of an advertisement from the time. One wonders how different the world would be today if

Microsoft had pushed Xenix instead of its own MS-DOS, and if AT&T had been easier to deal with than it apparently was. According to The Unix Heritage Society, the Santa Cruz Operation (SCO) later acquired Xenix, which in the mid to late 1980s was the most common Unix variant as measured by the number of machines on which it was installed.

Figure 8.2: Xenix: Microsoft's version of Unix

8.2 Unix wars

In the late 1980s there were numerous vendors of Unix systems, all using the trademarked name, and purveying software at least originally based on Version 7 from Bell Labs research. There were incompatibilities, however, particularly between AT&T's System V and the Berkeley distributions. All parties agreed that it would be highly desirable to have a common standard, but naturally disagreed on what it would be.

X/Open, an industry consortium, was formed in 1984 to try to create a standard source-code environment so that programs could be compiled on any Unix system without change.

AT&T and some allies formed their own group, Unix International, to promulgate their standard, in competition with a different standard from a group called the Open Software Foundation. The result? Two competing and different "open" standards. Eventually peace broke out, with a standard called POSIX (Portable Operating System Interface) for basic library functions, and a "Single Unix Specification" administered by X/Open that between them standardized libraries, system calls, and a large number of common commands (including the shell, Awk, `ed`, and `vi`).

In 1992, USL and AT&T sued Berkeley over intellectual property rights in Unix, claiming that Berkeley was using AT&T code without permission. Berkeley had made many changes in the AT&T code, and added much valuable material of their own, including the TCP/IP code that made the Internet accessible.

Berkeley continued to remove and rewrite code that had come from AT&T, and in 1991 released a version of Unix that they felt contained none of AT&T's proprietary material. AT&T and USL were not convinced, however, and this led to the lawsuit. After much maneuvering the case was heard in a New Jersey court, where Berkeley prevailed, in part on the grounds that AT&T had failed to put proper copyright notices on the code that it had distributed. Counter-suits ensued.

If this all sounds terminally complicated and boring, you're right, but it was a big deal at the time, with much waste of time and money on all sides. In 1991, AT&T sold shares in USL to 11 companies, and in 1993 Novell bought the rights to USL and Unix. Novell's CEO, Ray Noorda, decided to settle any remaining legal cases, perhaps realizing that the parties involved were spending more money on lawyers than they could possibly ever recoup in sales.

In retrospect, I suppose one could say that all the legal wrangling was a by-product of AT&T's early and almost accidental decision to make Unix available to universities. As Unix spread from universities using it for free to companies that were willing to pay for it, it became commercially viable, at least potentially. But it was too late for effective protection. Even if AT&T's source code was restricted, the system call interface was in effect in the public domain, and there was so much expertise in the community that creating versions unencumbered by AT&T licenses was almost routine. The same was true of application software like compilers, editors, and all the tools. To mix several metaphors, AT&T was trying to lock the barn door after the crown jewels had flown the coop.

8.3 Minix and Linux

AT&T's licensing of Unix became more and more restrictive as the company tried to make money from the software. This included restrictions on how Unix could be used in universities, which gave an advantage to BSD, which had no such constraints. At the same time, the ongoing wars between AT&T and BSD encouraged others to try rolling their own Unix-like systems. Independently created versions were free of commercial restrictions, since they used only the system call interface, but no one else's code.

At the Free University in Amsterdam in 1987, Andy Tanenbaum created Minix, a Unix lookalike that was compatible at the system call level but written entirely from scratch with a different kernel organization.

Minix was comparatively small, and to help with its spread, Andy wrote a textbook on it that metaphorically was a parallel to the Lions book of a decade earlier. Minix source code was available for free—one edition of the book came with a set of about a dozen floppy disks that could be loaded onto an IBM PC to provide a running system. I still have the first edition of Andy's book, and I might even have my Minix floppies as well.

Minix is alive and well today, a vehicle for education and for experimenting with operating systems.

One other result of AT&T's restrictive licensing, combined with the availability of Minix as a guide, was the independent development of another Unix-like system, compatible at the system call level, by a 21-year-old Finnish college student. On August 25, 1991, Linus Torvalds posted an item on comp.os.minix, a Usenet news group, shown in Figure 8.3.

Linus did not predict the remarkable future of his hobby system, any more than Ken and Dennis predicted the success of Unix. What started out as a few thousand lines of code now stands at well over 20 million lines, with Torvalds (Figure 8.4) as the principal developer and coordinator of a worldwide developer community maintaining and enhancing it. Torvalds is also the creator of Git, the most widely used version-control system for tracking code changes in software systems, of course including Linux.

At this point, Linux is a commodity operating system that can run on any kind of computer. It's on literally billions of devices (all Android phones, for example). It runs a substantial part of Internet infrastructure, including servers for major operations like Google, Facebook, Amazon, and so on. It's inside many Internet of Things (IoT) devices—my car runs Linux, as does my TV, your Alexa and Kindle, and your Nest thermostat. At the other end of the computing horsepower spectrum, it's the operating system on all of the top 500 supercomputers in the world. It's not a significant player in markets

```
Hello everybody out there using minix -

I'm doing a (free) operating system (just a hobby, won't be
big and professional like gnu) for 386(486) AT clones. This
has been brewing since april, and is starting to get ready.
I'd like any feedback on things people like/dislike in
minix, as my OS resembles it somewhat (same physical layout
of the file-system (due to practical reasons) among other
things).

I've currently ported bash(1.08) and gcc(1.40), and things
seem to work. This implies that I'll get something
practical within a few months, and I'd like to know what
features most people would want. Any suggestions are
welcome, but I won't promise I'll implement them :-)

              Linus (torv...@kruuna.helsinki.fi)

PS. Yes - it's free of any minix code, and it has a
multi-threaded fs. It is NOT protable (uses 386 task
switching etc), and it probably never will support anything
other than AT-harddisks, as that's all I have :-(.
```

Figure 8.3: Linus Torvald's announcement of Linux, August 1991

like laptop and desktop computers, however; there the majority run Windows, followed by macOS.

As a modern aside, the issue of whether it's possible to copyright a programming interface like the C standard library or the system calls for an operating system is now the center of an apparently endless lawsuit between Oracle and Google. Oracle acquired Sun Microsystems in 2010 and thus became the owner of the Java programming language. Later that year, it sued Google, claiming that Google was using Oracle's copyrighted Java interface in Android phones without permission, along with some patent claims. Google won this case, with the judge ruling that the patent claims were invalid and the Java API (application programming interface) could not be copyrighted.

Oracle appealed, and eventually a new trial was held. Google won again, but Oracle filed another appeal and this time the appeals court ruled for Oracle. Google appealed to the Supreme Court for the right to present the case there, in the hope of resolving the issue of whether APIs (not implementations!) could be copyrighted and thus used to prevent other parties from using an interface specification to create lookalike systems.

Figure 8.4: Linus Torvalds in 2014 (Wikipedia)

Disclosure: I've signed on to a couple of amicus briefs on the Google side here, since I do not believe that APIs should be copyrightable. If they were, we would not have had any of the Unix lookalikes, including Linux, since they are based on independent implementation of the Unix system call interface. We would probably also not have packages like Cygwin, a Windows implementation of Unix utilities that provides a Unix-like commandline interface for Windows users. Indeed, we would not likely have many independent implementations of interfaces for anything if they could be restricted by companies that claimed ownership.

At the time I'm writing this, the Supreme Court has not decided whether to hear the case. We shall see, since once the Court decides, that's it, short of Congress changing the law in a clear fashion. And of course who knows what might happen in other countries.

8.4 Plan 9

By the mid to late 1980's, Unix research in 1127 was slowing. The 7th edition, which was widely distributed and formed the base for most external versions, had been released in 1979. The 8th edition appeared six years later in 1985, the 9th in 1986, and the 10th, the final research version, was completed in 1989, though it was not distributed externally.

The perception was that Unix was a mature commercial system, and no longer a suitable vehicle for research in operating systems. A small group—Ken Thompson, Rob Pike, Dave Presotto and Howard Trickey—gathered together to work on a new operating system, which they called Plan 9 from Bell Labs after the 1959 science-fiction movie *Plan 9 from Outer Space*. (*Plan 9* the movie had gained a reputation as the worst movie ever made—a tough competition, to be sure—though some movie fans felt that it was so bad that it was actually good in a perverse way.)

Plan 9 the operating system was in part an attempt to take the good ideas in Unix and push them further. For example, in Unix, devices were files in the file system. In Plan 9, many more data sources and sinks were files as well, including processes, network connections, window-system screens and shell environments. Plan 9 also aimed for portability right from the beginning, with a single source that could be compiled for any supported architecture. Another outstanding feature of Plan 9 was its support for distributed systems. Processes and files on unrelated systems with different architectures could work together exactly as if they were in the same system.

Plan 9 was made available to universities in 1992, and released publicly a few years later for commercial use, but aside from a small community of aficionados, it is not used much today. The major reason is probably that Unix and increasingly Linux simply had too much momentum and there wasn't a compelling reason for most people to switch systems. A smaller part of the reason it didn't catch on may be that it took a rather Procrustean view. Plan 9 mechanisms were in many cases better than the Unix equivalents, and there wasn't much attempt to provide compatibility. For example, Plan 9 originally did not provide the C standard I/O library stdio, instead using a new library called *bio*. *Bio* was cleaner and more regular than stdio, but without the standard library, it took real work to convert programs to run on both Unix and Plan 9. Similarly, there was a new version of Make called Mk, which was superior in many ways, but it was incompatible and so existing makefiles had to be rewritten.

There were mechanisms for conversion, and Howard Trickey (Figure 8.5) ported a number of key libraries like stdio, but in spite of that, at least for some potential users, including me, it was too much effort. Thus Plan 9 was unable to profit directly from a lot of good Unix software, and it was harder to export its software innovations to the mainstream Unix world.

Plan 9 did contribute one thing of surpassing importance to the world, however: the UTF-8 encoding of Unicode.

Unicode is an ongoing effort to provide a single standard encoding of all the myriad characters that mankind has ever used for writing. That includes

Figure 8.5: Howard Trickey, ~1984 (Courtesy of Gerard Holzmann)

alphabetic scripts like those in most Western languages, but also ideographic scripts like Chinese, ancient scripts like Cuneiform, special characters and symbols of all types, and recent inventions like emojis. Unicode has nearly 140,000 characters at the moment, and the count increases slowly but steadily.

Unicode was originally a 16-bit character set, big enough to hold all alphabetic scripts and the roughly 30,000 characters of Chinese and Japanese. But it was not feasible to convert the world to a 16-bit character set when most computer text was in ASCII, a 7-bit set.

Ken Thompson and Rob Pike wrestled with this issue for Plan 9, since they had decided that Plan 9 would use Unicode throughout, not ASCII. In September 1992, they came up with UTF-8, a clever variable-length encoding of Unicode. UTF-8 is efficient in both space and processing time. It represents each ASCII character as a single byte, and uses only two or three bytes for most other characters, with a maximum of four bytes. The encoding is compact, and ASCII is legal UTF-8. UTF-8 can be decoded as it is read, since no legal character is a prefix of any other character, nor is any character part of any other character or sequence of characters. Almost all text on the Internet today is encoded in UTF-8; it is used everywhere by everyone.

8.5 Diaspora

In 1996, AT&T split itself again, this time voluntarily and into three parts, a process that called for a new word: "trivestiture." One part remained AT&T, focusing on long distance telephony and communications. A second part became Lucent Technologies, which was in effect the follow-on to Western Electric, focused on manufacturing telecom equipment. (One of the company's slogans was "We make the things that make communications work.") The third fission product came from undoing the ill-judged acquisition of NCR in 1991 when AT&T was trying to enter the computer business.

Bell Labs employees at the time were largely skeptical of trivestiture. The hype that came with the new company name and logo was met with some scorn. Figure 8.6 shows the flaming red Lucent logo that was announced with much fanfare in 1996; I can't repeat most of the names that were soon attached to it. Figure 8.7 is an on-point Dilbert cartoon that appeared soon afterwards.

Figure 8.6: Lucent Technologies logo

Figure 8.7: Dilbert's take on the Lucent logo? DILBERT © 1996 Scott Adams. Used by permission of ANDREWS MCMEEL SYNDICATION. All rights reserved.

Trivestiture split Bell Labs Research itself along functional lines, with roughly one third of the research staff going to AT&T to form AT&T Labs (now AT&T Shannon Labs), and the rest remaining "Bell Labs" as part of Lucent. For the most part, people went where they were told, but the members of 1127, with a long history of pushing back against management edicts, strongly resented the forced split-up of the Center. We took a hard line, and management reluctantly agreed to let everyone make their own choice. In a stressful real-time process, each person decided whether to go with AT&T or stay with Lucent. In the end, this achieved the same roughly one-third / two-thirds split that had been planned, but individuals had control over their own destinies, at least in the short run.

All parties ultimately fell on hard times. AT&T was eventually bought by Southwestern Bell (now SBC Communications), one of the original Baby Bells. SBC rebranded itself with the AT&T name, logo and even the stock ticker symbol "T" that had been used as early as 1901.

Lucent went through a boom and then a bust, with some questionable business practices en route. As it struggled to survive, it spun off its enterprise communications services business into a company called Avaya in 2000, and its integrated-circuit business into another called Agere in 2002. Each split removed more people from Bell Labs, narrowing the breadth of research activities and of course shrinking the financial base that could support long-term work. Agere was eventually absorbed into LSI Logic; after some significant ups and downs, including bankruptcy, Avaya is still in business as an independent company.

In 2006 Lucent merged with the French telecom company Alcatel to form Alcatel-Lucent, which in turn was taken over by Nokia in 2016. Bell Labs was swept up in this wave of mergers and takeovers, but along the way, most of those who had been involved with Unix and Center 1127 dispersed to other places. The number 1127 itself disappeared in a reorganization in 2005.

Gerard Holzmann maintains a list of the alumni of Center 1127 and where they are, at *www.spinroot.com/gerard/1127_alumni.html*. All too many have died, but of those still alive, the most common destination has been Google; others are at other companies, teaching, or retired. Only a tiny handful remain at Bell Labs.

Chapter 9

Legacy

"Unix was not only an improvement on its predecessors but also on most of its successors."
 Doug McIlroy, *Remarks for Japan Prize award ceremony for Dennis Ritchie*, May 2011, paraphrasing Tony Hoare on Algol

Unix has been tremendously successful. As Unix or Linux or macOS or other variants, it runs on billions of computers, serves billions of people continuously, and has of course made billions of dollars for any number of people who have built on top of it (though not for any of its creators). Later operating systems have been strongly shaped by its decisions.

Languages and tools originally developed at Bell Labs for Unix are everywhere. Among the programming languages are C and C++, which are still the backbone of system programming today and more specialized ones like Awk and AMPL. Core tools include the shell, diff, grep, Make and Yacc.

GNU (a recursive acronym for "GNU's not Unix") is a large collection of software, much based on Unix models, that is freely available in source-code form for anyone to use: it makes almost everything from Unix available, along with much more. Coupled with the Linux operating system, GNU amounts to a free version of Unix. GNU implementations of Unix commands are open source, so they can be used and extended, subject only to the restriction that if the improvements are distributed, they have to be made available to everyone for free; they can't be taken private. A huge amount of today's software development is based on open source and in many cases GNU implementations.

What accounts for the success of Unix? Are there ideas or lessons that can be learned and applied in other settings? I think that the answer is yes, on at least two fronts: technical for sure, and organizational as well.

9.1 Technical

The important technical ideas from Unix have been discussed in the first few chapters of the book; this section is a brief summary. Not everything here originated with Unix, of course; part of the genius of Ken Thompson and Dennis Ritchie was their tasteful selection of existing good ideas, and their ability to see a general concept or a unifying theme that simplified software systems. People sometimes talk of software productivity in terms of the number of lines of code written; in the Unix world, productivity was often measured by the number of special cases or lines of code that were removed.

The *hierarchical file system* was a major simplification of existing practice, though in hindsight it seems utterly obvious—what else would you want? Rather than different types of files with properties managed by the operating system, and arbitrary limits on the depth to which files could be nested in directories, the Unix file system provides a straightforward view: starting from the root directory, each directory contains either information about files, or directories that contain information about further files and directories. Filenames are simply the path from the root, with components separated by slashes.

Files contain uninterpreted bytes; the system itself does not care what the bytes are, or know anything about their meaning.

Files are created, read, written and removed with half a dozen straightforward system calls. A handful of bits define access controls that are adequate for most purposes. Entire storage devices like removable disks can be mounted on a file system so that logically the contents become part of the file system at that point.

Naturally, there are some irregularities. Devices appear in the file system, which is a simplification, but operations on them, especially terminals, have special cases and an interface that remains messy even today.

What I've described here is the logical structure of the file system. There are plenty of ways to implement this model, and in fact modern systems support a wide variety of implementations, all presenting the same interface but with different code and internal data structures to make it work. If you look at your computer, you will see multiple devices that use this model: the hard drive, USB thumb drives, SD cards, cameras, phones, and so on. The brilliance of Unix was in choosing an abstraction that was general enough to be remarkably useful, yet not too costly in performance.

A *high-level implementation language*, for user programs of course, but also for the operating system itself. The idea was not new; it had been tried in Multics and a couple of earlier systems, but the time and the languages were not quite ready. C was much more suitable than its predecessors, and it led to portability of the operating system. Where once there were only proprietary operating systems from hardware manufacturers, often with their own proprietary languages, Unix became an open and widely understood standard and then a commodity: the system could be used on all computers with only minor changes. Customers were no longer locked in to specific hardware, and manufacturers no longer had to develop their own operating systems or languages.

The *user-level programmable shell*, with control-flow statements and easy I/O redirection, made it possible to program by using programs as building blocks. As the shell's programming capabilities grew, it became another high-level language in the programmer's toolbox. And because it was a user-level program, not part of the operating system, it could be improved on and replaced by anyone with a better idea. The evolution from the original Unix shell through PWB, the Bourne shell and Bill Joy's `csh` to today's proliferation illustrates the benefits, and of course some of the drawbacks—it's all too easy for incompatible versions to multiply.

Pipes are the quintessential Unix invention, an elegant and efficient way to use temporary connections of programs. The notion of streaming data through a sequence of processing steps is natural and intuitive, the syntax is exceptionally simple, and the pipe mechanism fits perfectly with the collection of small tools. Pipes do not solve all connection problems, of course, but the fully general non-linear connections of Doug McIlroy's original concept do not show up often in practice; linear pipelines are almost always good enough.

The notion of *programs as tools* and using them in composition is characteristic of Unix. Writing small programs that each do one thing well, rather than large and monolithic programs that try to do many things, has many benefits. Certainly there are times when monoliths make sense, but there are strong advantages to a collection of small(ish) programs that ordinary users can combine in novel ways.

In effect, this approach is modularization at the level of whole programs, parallel to modularization at the level of functions within a program. In both cases, the approach amounts to a kind of divide and conquer, since the individual components are smaller and don't interact with each other. It also permits a mix-and-match functionality that is hard to achieve with big programs that try to do too many different things in a single package.

Ordinary text is the standard data format. The pervasive use of text turns out to be a great simplification. Programs read bytes, and if they are intended to process text, those bytes will be in a standard representation, generally lines of varying length, each terminated with a newline character. That doesn't work for everything, but pretty much anything can use that representation with little penalty in space or time. As a result, all those small tools can work on any data, individually or in combination.

It's interesting to speculate about how differently things might have turned out if Unix had been developed in a world with punch cards instead of Teletypes. Punch cards practically force a world-view that everything comes in 80-character chunks, with information most often located in fixed fields within the chunks.

Programs that write programs is a powerful idea. Much of the progress that we have made in computing has been through mechanization—getting the computer to do more of the work for us. It's hard to write programs by hand, so if you can get a program to write them for you, it's a big win. It's way easier, and the generated programs are more likely to be correct.

Compilers are of course an old example, but at a higher level, Yacc and Lex are excellent examples of generating code to create programming languages. Tools for automation and mechanization, like shell scripts and makefiles, are in effect programs that create programs. These tools are still widely used today, sometimes in the form of very large configuration scripts and makefile generators that accompany the source code distributions of languages like Python and compilers like GCC.

Specialized languages, today often called little languages, domain-specific languages, or application-specific languages. We tell computers what to do by using languages. For most programmers, that means general-purpose languages like C, but there are a host of more specialized languages that focus on narrower domains.

The shell is a good example: it's meant for running programs, and it's very good at that, but you would not want to use it to write a browser or a video game. Specialization is an old idea, of course; the earliest high-level languages aimed at specific targets, for instance Fortran at scientific and engineering computation, and Cobol at business data processing. Languages that tried too early for too broad a range of applicability have sometimes foundered; PL/I is an instance.

Unix has a long tradition of special-purpose languages, well beyond the shell. The document preparation tools that are close to my heart are good examples, but so are calculators, circuit design languages, scripting languages, and the ubiquitous regular expressions. One of the reasons there were so many languages is the development of tools that enabled non-experts to create them. Yacc and Lex are the primary examples here, and they are specialized languages in their own right.

Of course, not all languages need to be high-tech to be useful. Steve Johnson built the first version of the `at` command in an evening:

> "Unix had a way of scheduling jobs to be run 'out of hours' so that long jobs wouldn't interfere with peoples' work (remember, there were a dozen or so people sharing the Unix machine). To cause a job to run later, you needed to edit a system file and fill in a table of information in a rather obscure format. I was attempting this feat one day and heard myself muttering 'I want this job to run at 2AM.' Suddenly, I realized that the information about what to do could be captured in a simple syntax: 'at 2AM run_this_command.' I hacked together a version in a couple of hours and advertised it in the 'message of the day' file the next morning."

The `at` command is still used 40-plus years later, with not many changes. Like a number of other languages, the syntax is a sort of stylized English based on how we say things aloud.

The Unix philosophy, a style of programming, of how to approach a computing task, was summarized by Doug McIlroy in his foreword to the special issue of the *Bell Labs Technical Journal* on Unix, in July 1978:

> (i) Make each program do one thing well. To do a new job, build afresh rather than complicate old programs by adding new "features."

> (ii) Expect the output of every program to become the input to another, as yet unknown, program. Don't clutter output with extraneous information. Avoid stringently columnar or binary input formats. Don't insist on interactive input.

> (iii) Design and build software, even operating systems, to be tried early, ideally within weeks. Don't hesitate to throw away the clumsy parts and rebuild them.

> (iv) Use tools in preference to unskilled help to lighten a programming task, even if you have to detour to build the tools and expect to throw some of them out after you've finished using them.

These are maxims to program by, though not always observed. One example: the `cat` command that I mentioned in Chapter 3. That command did one thing, copy input files or the standard input to the standard output. Today the GNU version of `cat` has (and I am not making this up) 12 options, for tasks like numbering lines, displaying non-printing characters, and removing duplicate blank lines. All of those are easily handled with existing programs; none has anything to do with the core task of copying bytes, and it seems counter-productive to complicate a fundamental tool.

The Unix philosophy certainly doesn't solve all the problems of programming, but it does provide a useful guide for approaching system design and implementation.

9.2 Organization

I believe that another large component of the success of Unix was due to non-technical factors, like the managerial and organizational structure of Bell Labs, the social environment of 1127, and the flow of ideas across a group of talented people working on diverse problems in a collegial environment. These are harder to assess than technical notions, so this is necessarily a more subjective view. As with the previous section, most of these have been mentioned earlier.

A *stable environment* is crucial: money, resources, mission, structure, management, culture—all should be consistent and predictable. As described in Chapter 1, Bell Labs research was a large operation inside a large development organization within a large corporation with a long history and a clear mission: universal service. The long-term Bell Labs goal of continuously improving telephone service meant that researchers could explore ideas that they thought were important, for long periods, even years, without having to justify their efforts every few months. There was oversight, of course, and anyone who worked on a project for several years without producing anything would be encouraged to change something. Occasionally, someone was eased out of research or out of the company entirely, but in my 15 years in management, I remember only a handful of cases.

Funding was assured, and working-level researchers did not have to think about money, nor did it concern me even when I was a department head. Certainly at some level someone did worry about such matters, but not the people doing the research. There were no research proposals, no quarterly progress reports, and no need to seek management approval before working on something. Somewhere into my time as department head, I did start to have to generate semi-annual reports on activities in my department, and for that I solicited a paragraph from each person. These were gathered for information only, however, not for evaluating performance. There were occasional periods when travel was scrutinized more carefully—we might be limited to one or two conferences a year, perhaps—but for the most part, if we needed to buy equipment or make a trip, money was available without much question.

Problem-rich environment. As Dick Hamming said, if you don't work on important problems, it's unlikely that you'll do important work. But almost any topic had the potential to be important and relevant to AT&T's communications mission. Computer science was a new field, with plenty of ideas to explore on both theoretical and practical fronts, and of course the interplay between theory and practice was particularly fruitful. Language tools and regular expressions are good examples.

Within AT&T, the use of computers was exploding, and Unix was a big part of that, especially in systems for operations support, as seen with the Programmer's Workbench. The mainline telephone business was changing too, as electro-mechanical telephone switches gave way to computer-controlled electronic switching. Again, this was a source of interesting data and projects to work on, and often to contribute as well. One downside: most of the switching work was done at very large development divisions in Indian Hill (Naperville, Illinois), so collaboration often required trips to Chicago. Distance is hard to overcome, and is still an issue today. Even with excellent video conferencing, there is no substitute for having your collaborators next door and your stable of experts nearby.

Bell Labs scientists were also expected to be part of the academic research community, another source of research problems and insights, and to keep up with what was going on in other industrial research labs like Xerox PARC and IBM Watson. We attended the same conferences, published in the same journals, and often collaborated with academic colleagues, with sabbaticals in both directions. For example, I spent the fall of 1996 teaching at Harvard, with the full support of Bell Labs; they even paid my salary, so Harvard got a freebie. The same was true in the academic year 1999-2000 at Princeton.

In addition to internal courses, any number of colleagues taught at universities. It was easily arranged for nearby schools like Princeton, NYU, Columbia and West Point, and not much harder for extended visits to places further away: Ken Thompson spent a year at Berkeley, Rob Pike spent a year in Australia, Doug McIlroy spent a year at Oxford. External visibility was important for recruiting as well as for generally keeping up with the field. Secretive companies had a harder time attracting talent, something that appears to still be true today.

Hire the best. One resource was very carefully managed: hiring. In 1127, we typically could only hire one or two people a year, and almost always young ones, so hiring decisions were made very cautiously, perhaps too much so. This is of course a familiar problem in university departments as well. It is often unclear whether to go after a star in a particular field, or someone else who is broadly talented; as Steve Johnson once put it, should we be hiring athletes or first basemen? My preference has been for people who are really good at what they do, without worrying too much about specifically what it is.

In any case, Bell Labs worked hard to try to attract good people. Recruiters from Research visited major computer science departments once or twice a year, looking at PhD candidates. When someone promising was identified, he or she was invited to visit for a couple of days, and would normally be interviewed by several groups, not just 1127. The programs like GRPW and CRFP, for women and minorities, that I mentioned in Chapter 1 were a big help here as well, since they provided first-rate candidates for permanent employment who had already spent significant time with us during their graduate school years.

We used our own researchers as recruiters, not professional career recruiters. Someone doing active research of their own could have technical conversations with faculty and students, would always learn something useful, and could leave a positive impression of the company.

Relationships with universities tended to be long term. I was the recruiter for Carnegie-Mellon in Pittsburgh for at least 15 years. Twice a year I would spend several days at CMU, talking with computer science faculty about their research and with students who might be interested in working at Bell Labs. I made good friends in this way even if they didn't join the Labs. Recruiting was very competitive, since good universities were hiring actively, as were top industrial research labs, so my personal list of the ones who got away is long. I was certainly right to want to recruit most of them; they've been a highly successful group.

Technical management. Managers have to understand the work they manage. Management in Bell Labs research was technical all the way to the top, so they had a solid understanding of the work both within their own organizations and across others. Department heads were expected to know what their people were doing in real detail, not for the purpose of arguing how great it was, but for being able to explain it to others and helping to make connections. At least in 1127 there were no turf wars; it was a cooperative and non-competitive environment where management supported their people, often collaborated, and never competed. I'm not sure that this was a universal experience, but it's worth aiming for, and doing it well should be part of the reward mechanism for managers.

Although Bell Labs management was technically knowledgeable at all levels, AT&T's upper management seemed removed from new technology and slow to adapt to change. For instance, in the early 1990s, Sandy Fraser, at that time the director of 1127, told AT&T executives that networking improvements would mean that long distance prices would come down from their then-current ten cents per minute to one cent a minute, and he was laughed at. Today's price is pretty close to zero cents per minute; Sandy was too conservative.

Cooperative environment. The size and scale of Bell Labs meant that for almost any technical area, there were multiple experts, often world leaders in their field. Furthermore the culture was strongly cooperative and helpful. It was absolutely standard procedure to walk into someone's office and ask for help; most often the person being asked would drop everything to assist. There was also a superb technical library, open 24 hours a day, with subscriptions to a large collection of journals, and remote access to other libraries; it was equivalent to a university library but focused on science and technology.

For many people in 1127, the closest collection of relevant experts was the Mathematics Research Center, 1121, which had extraordinary mathematicians, including Ron Graham, Mike Garey, David Johnson, Neil Sloane, Peter Shor, Andrew Odlyzko—the list just goes on. John Tukey, arguably the world's foremost statistician at the time (and incidentally the person who coined the word "bit"), was just across the hall, and there were formidable experts on pretty much any aspect of mathematics and communications. For instance, my current Princeton colleague Bob Tarjan, who shared the Turing Award in 1986, was in the math center.

They were always ready to help, and not always just on technical matters. For example, in addition to being an outstanding mathematician, Ron Graham was an expert juggler and a former president of the International Jugglers' Association; he even had a net in his office to catch dropped juggling balls before they hit the floor. Ron used to say that he could teach anyone to juggle in 20 minutes. I fear that it wasn't true in my case, but an hour of one-on-one instruction (in his office!) did get me over the hump, and I still have the lacrosse balls that he gave me to practice with.

Fun. It's important to enjoy your work and the colleagues that you work with. 1127 was almost always a fun place to be, not just for the work, but the esprit of being part of a remarkable group. Since there were no local options other than the company cafeteria, lunchtime provided a mix of social and technical discussions. While the Unix room crowd normally ate at 1PM, a larger group routinely ate at 11AM. Topics included everything from technical ideas both big and small to politics with no holds barred; these would often be continued during a walk around the Bell Labs property.

Center members played pranks on each other, and took perhaps undue pleasure from pushing back against the bureaucracy that's inevitable in any big company. I've already mentioned disdain for badges. The various forms and procedures that we were supposed to use provided further opportunities.

For example, security staff would ticket cars that were violating some or other rule. One spring day, Mike Lesk found a blank ticket, which he put on the windshield of a colleague's car with the violation listed as "Failure to remove ski-rack by April 1." The colleague, not to be identified here, was actually taken in by this for a few hours.

By far the most elaborate prank was played on Arno Penzias by a team of at least a dozen, led by Rob Pike and Dennis Ritchie, with the aid of professional magicians Penn and Teller. It's too long for the book, but Dennis tells the Labscam story at *www.bell-labs.com/usr/dmr/www/labscam.html*, and the video is at *www.youtube.com/watch?v=if9YpJZacGI*. I'm in the credits at the end as a gaffer, which is accurate—much duct tape was involved.

There was no free food at Bell Labs (that's a modern perk that I would have appreciated back in the day), but somehow we managed free coffee; management quietly paid for it.

People would leave offerings in the Unix room for the common good. Someone found a supply of 10 kg (22 lb) blocks of high-quality chocolate and left them for people to chip away at it. The food wasn't always up to that standard, however:

> "Somebody brought in a bag of objects labeled in Chinese. All of us bit into one of them and gave up. We then noticed that they were disappearing: it turned out that [redacted] was eating them. Near the end of the bag somebody who knew Chinese told us that the instructions on the bag said to soak for an hour in boiling water before eating."

At the same time, there was zero, or even negative, enthusiasm for the kinds of team-building exercises that one often sees today. Most of us saw them as artificial, pointless, and a waste of time.

It takes effort to build and maintain an organization whose members like and respect each other, and who enjoy each other's company. This can't be created by management fiat, nor by external consultants; it grows organically from the enjoyment of working together, sometimes playing together, and appreciating what others do well.

9.3 Recognition

Unix and its primary creators, Ken Thompson and Dennis Ritchie, have been recognized for their contributions. When they won the ACM Turing Award in 1983, the award selection committee said

> "The success of the Unix system stems from its tasteful selection of a few key ideas and their elegant implementation. The model of the Unix system has led a generation of software designers to new ways of thinking about programming. The genius of the Unix system is its framework, which enables programmers to stand on the work of others."

They also received the US National Medal of Technology in 1999. Figure 9.1 is a picture with President Bill Clinton, one of the few occasions where either Ken or Dennis wore a suit and tie. Bell Labs was for us at least a very informal environment by the standards of the time. As Dennis said in his online biography, "Ken's virtual coat-tails are long. I'm one of the few, besides Bonnie T., who has seen him wear a real coat (and even black tie) on more than one occasion." I personally have never seen Ken dressed up at all.

Other awards include membership in the National Academy of Engineering, and the Japan Prize in Information and Communications in 2011, for which the citation reads

> "Compared to other operating systems prevailing around that time, their new and advanced OS was simpler, faster and featured a user-friendly hierarchical file system. Unix was developed in conjunction with the programming language, C, which is still widely used for writing OS, and dramatically improved the readability and portability of Unix source code. As a result, Unix has come to be used by various systems such as embedded systems, personal computers, and super computers."

> "Unix was also a major driving force behind the development of the Internet. University of California, Berkeley developed Berkeley Software Distribution (BSD), an extended version of Unix that was implemented with the Internet protocol suite TCP/IP. The development was based on the sixth edition of Unix that Bell Labs distributed along with its source code to universities and research institutions in 1975, which led to the beginning of an 'open source' culture. BSD Unix helped the realization of the Internet."

Figure 9.1: Ken, Dennis, Bill, National Medal of Technology, 1999

Other forms of recognition were more informal, signs that Unix and C had entered popular culture, like the appearance of C on the popular TV program *Jeopardy*:

```
From dmr@cs.bell-labs.com Tue Jan  7 02:25:44 2003
Subject: in case you didn't see it

On Friday night on "Jeopardy!", in a category called
"Letter Perfect" (all the answers were single letters),
the $2,000 (most difficult) question was:

DEVELOPED IN THE EARLY 1970S, IT'S THE MAIN PROGRAMMING
LANGUAGE OF THE UNIX OPERATING SYSTEM.
```

and in some kind of high point for geeks, a famous scene in the 1993 movie *Jurassic Park*, where the 13-year-old Lex Murphy (Ariana Richards) says "It's a Unix system! I know this." She navigates the file system, finds the controls for the doors, locks them, and thus saves everyone from being eaten by velociraptors (Figure 9.2).

Other members of Center 1127, in part thanks to the rich environment enabled by Unix, have also been recognized professionally, like the 8 other 1127 alumni who are members of the National Academy of Engineering.

Figure 9.2: Unix in *Jurassic Park*

9.4 Could history repeat?

Could there be another Unix? Could a new operating system come out of nowhere and take over the world in a few decades? I often get this question when I talk about Unix. My answer is no, at least at the moment. There will be no revolution; more likely, operating systems will continue to evolve, while carrying a great deal of Unix DNA.

But some analogous success could happen in other areas of computing. There are always creative people, good management is not unheard of, hardware is very cheap, and great software is often free. On the other hand, there are few unfettered environments, industrial research is much reduced and constrained and far more short-term than it was fifty years ago, and academic research is always strapped for funds.

Nevertheless, I am optimistic, on the grounds that the great ideas come from individuals.

For example, the number of people contributing to Unix in the early days was tiny; arguably the core was a single person, Ken Thompson, who is certainly the best programmer I have ever met, and an original thinker without peer. Dennis Ritchie, whose name is linked with Ken as the co-creator of Unix, was a vital contributor, and his C programming language, central to the evolution of Unix in the early days, is still the *lingua franca* of computing.

It's instructive to examine the languages that programmers use every day and see how often they were originally the work of one or two people. Almost every major programming language is like that, including Java (James Gosling), C++ (Bjarne Stroustrup), Perl (Larry Wall), Python (Guido von Rossum), and JavaScript (Brendan Eich). It seems safe to predict that there will continue to be new languages to make programming easier and safer. It's also safe to predict that there won't be just one language, however—there are too many tradeoffs for a single language to serve all purposes well enough.

Google, Facebook, Amazon, Twitter, Uber, and any number of other companies that went from startups to multi-billion-dollar enterprises originated with a bright idea by one or two people. There will be more of these, though it's also possible that as new ideas occur and new companies appear, they will quickly be snapped up by the existing big companies. The bright ideas may be preserved, and the inventors will be well compensated, but the big fish are likely to eat the little ones quite quickly.

Good management is another component of success. Doug McIlroy stands out here as unique, a leader of outstanding intellectual horsepower, with incomparable technical judgment, and a management style based on being one of the first users of whatever his colleagues developed. Unix itself, but also languages like C and C++, and any number of Unix tools, all profited from Doug's good taste and razor-sharp critiques. So did Unix documentation of all sorts, from the user manuals through a few dozen influential books. I can attest to this personally: Doug was the outside reader on my PhD thesis in 1968, made incisive comments on all of my technical papers and books, and is still keeping me on target more than 50 years later.

Bell Labs management was technically competent, and especially so in 1127, so it could assess good work, and it was hands-off, so it didn't push particular projects or approaches. In over 30 years at the Labs I was never told what to work on. Bruce Hannay, vice president of Research after Bill Baker, said in 1981 in *Engineering and Science in the Bell System*,

> "Freedom of choice is of utmost importance to the research scientist, because research is an exploration of the unknown and there are no road maps to tell what course to take. Each discovery affects the future course of research and no one can predict or schedule discovery. Thus Bell Laboratories research managers have provided the maximum possible freedom to the research staff, consistent with the institutional purpose. Research people have been chosen for their creative abilities and are encouraged to exercise these to the fullest."

One of the best examples I ever saw of this nearly absolute freedom was the work that Ken Thompson and Joe Condon did on their chess computer. One day Bill Baker, president of Bell Labs, brought some or other distinguished visitor to the Unix room, and Ken showed off Belle. The visitor asked why Bell Labs supported work on computer chess, since it didn't seem to have anything to do with telephones. Bill Baker answered: Belle was an experiment in special-purpose computers, it had led to the development of new circuit design and implementation tools, and it gave Bell Labs visibility in another field. Ken didn't have to do any justification whatsoever.

The big secret to doing good research is to hire good people, make sure there are interesting things for them to work on, take a long view, and then get out of the way. It certainly wasn't perfect, but Bell Labs research generally did this well.

Of course computing didn't exist in a technological vacuum. The invention of the transistor and then integrated circuits meant that for 50 years computing hardware kept getting smaller, faster, and cheaper at an exponential rate. As hardware got better, software became easier, and our understanding of how to create it got better as well. Unix rode the technology improvement wave, as did many other systems.

As I said in the Preface, Unix was probably a singularity, a unique combination of circumstances that changed the computing world. I doubt that we will see anything like it again in operating systems, but there will surely be other occasions when a handful of talented people with good ideas and a supportive environment do change the world with their inventions.

For me, Bell Labs and 1127 were a marvelous experience: a time and place with endless possibilities and a group of first-rate colleagues who made the most of them. Few are fortunate enough to have that kind of experience, the shared creation, of course, but especially the friends and colleagues with whom it was shared.

> "What we wanted to preserve was not just a good environment in which to do programming, but a system around which a fellowship could form. We knew from experience that the essence of communal computing [...] is not just to type programs into a terminal instead of a keypunch, but to encourage close communication."
>
> Dennis Ritchie, "The Evolution of the Unix Time-sharing System," October 1984

Sources

"The names of Ritchie and Thompson may safely be assumed to be attached to almost everything not otherwise attributed."

"To look further afield would require a tome, not a report, and possibly a more dispassionate scholar, not an intimate participant."
  Doug McIlroy, *A Research Unix Reader: Annotated Excerpts from the Programmer's Manual, 1971-1986*, 1986

Much Unix history is online (though not always in a searchable form), thanks to some good luck and truly dedicated work by amateur and professional historians, such as The Unix Heritage Society and the Computer History Museum. Further material is available through interview videos and oral histories; some are contemporaneous like the various AT&T public relations efforts, and some are retrospective. This list of sources is in no way complete or comprehensive, but it will give readers who want to dig further a good start. Many of these documents can be found on the Internet.

A History of Science and Engineering in the Bell System has seven volumes with nearly 5,000 pages written by members of technical staff at Bell Labs, mostly in the 1970s and 1980s. One volume deals with the relatively late advent of computing.

Bell Labs maintains a sequence of short pages on the history of Unix at *s3-us-west-2.amazonaws.com/belllabs-microsite-unixhistory*.

A. Michael Noll, a member of the Speech and Acoustics Research Center in the 1960s and early 1970s, has written a memoir of his time at the Labs, along with material from his role as the editor of the papers of Bill Baker; it can be found at *noll.uscannenberg.org* along with a variety of other informative historical information. It's an excellent read for basic facts about the

Labs and what it was like in the speech and acoustics research area. Mike's memories about the collegiality and openness of Bell Labs generally accord with mine, though he feels things started to fall apart much sooner than I do, perhaps because we were in different (though organizationally adjacent) areas.

Tom Van Vleck maintains a thorough repository of historical information about Multics at *multicians.org*.

The special issue of the *Bell System Technical Journal* on Unix in July 1978 has several fundamental papers, including an updated version of the CACM paper, Ken's "Unix Implementation," Dennis's "Retrospective," a paper by Steve Bourne on the shell, along with a paper on PWB by Ted Dolotta, Dick Haight and John Mashey.

The special issue of the *AT&T Bell Labs Technical Journal* on Unix in 1984 includes Dennis Ritchie's "Evolution of Unix" paper, and "Data Abstraction in C" by Bjarne Stroustrup, among other interesting articles.

Doug McIlroy's "A Research Unix Reader" is especially good for historical background; it can be found at *genius.cat-v.org/doug-mcilroy*.

The Unix Heritage Society, run by Warren Toomey with the help of many volunteers, has preserved versions of early Unix code and documentation; it's a great place to browse. For example, *www.tuhs.org/Archive/Distributions/Research/Dennis_v1* has the code for the First Edition as provided by Dennis Ritchie.

The late Michael Mahoney, professor of the History of Science at Princeton University, interviewed a dozen members of 1127 in the summer and fall of 1989 for an extensive oral history of Unix. Mike's raw transcripts and edited interviews are maintained by the History department at Princeton, and can be found at *www.princeton.edu/~hos/Mahoney/unixhistory*. In addition to being a first-rate historian, Mike was a programmer who really understood what his subjects were talking about, so there is often significant technical depth.

Phyllis Fox, a pioneer of numerical computing and of technical women at Bell Labs, did an oral history for the Society for Industrial and Applied Mathematics (SIAM) in 2005, available at *history.siam.org/oralhistories/fox.htm*; it includes a detailed description of the PORT portable Fortran libraries.

The May 2019 fireside chat with Ken Thompson at the Vintage Computer Festival East is on YouTube at *www.youtube.com/watch?v=EY6q5dv_B-o*.

Two books on the early history of Unix are available for free download: *Life with Unix* by Don Libes and Sandy Ressler (1989), and *A Quarter Century of Unix* by Peter Salus (1994).

Dennis Ritchie's home page at (Nokia) Bell Labs has been preserved. It has links to most of the papers that Dennis wrote and to other historical material. The link is *www.bell-labs.com/usr/dmr/www*.

Kirk McKusick, one of the central figures in BSD, has written a careful history of BSD, available at *www.oreilly.com/openbook/opensources/book/kirkmck.html*. Ian Darwin and Geoff Collyer provide additional insights from a somewhat different perspective in *doc.cat-v.org/unix/unix-before-berkeley*.

Printed in Great Britain
by Amazon